PROBABILITY THEORY

SYNTHESE LIBRARY

STUDIES IN EPISTEMOLOGY,

LOGIC, METHODOLOGY, AND PHILOSOPHY OF SCIENCE

Managing Editor:

JAAKKO HINTIKKA, *Boston University*

Editors:

DIRK VAN DALEN, *University of Utrecht, The Netherlands*
DONALD DAVIDSON, *University of California, Berkeley*
THEO A.F. KUIPERS, *University of Groningen, The Netherlands*
PATRICK SUPPES, *Stanford University, California*
JAN WOLEŃSKI, *Jagiellonian University, Kraków, Poland*

VOLUME 297

PROBABILITY THEORY

Philosophy, Recent History and Relations to Science

Edited by

VINCENT F. HENDRICKS

STIG ANDUR PEDERSEN

and

KLAUS FROVIN JØRGENSEN
*Department of Philosophy and Science Studies,
University of Roskilde, Denmark*

KLUWER ACADEMIC PUBLISHERS
DORDRECHT / BOSTON / LONDON

A C.I.P. Catalogue record for this book is available from the Library of Congress.

ISBN 0-7923-6952-1

Published by Kluwer Academic Publishers,
P.O. Box 17, 3300 AA Dordrecht, The Netherlands.

Sold and distributed in North, Central and South America
by Kluwer Academic Publishers,
101 Philip Drive, Norwell, MA 02061, U.S.A.

In all other countries, sold and distributed
by Kluwer Academic Publishers,
P.O. Box 322, 3300 AH Dordrecht, The Netherlands.

Printed on acid-free paper

All Rights Reserved
© 2001 Kluwer Academic Publishers
No part of the material protected by this copyright notice may be reproduced or
utilized in any form or by any means, electronic or mechanical,
including photocopying, recording or by any information storage and
retrieval system, without written permission from the copyright owner.

Printed in the Netherlands.

CONTENTS

Preface ix

Contributing Authors xi

V.F. HENDRICKS, S.A. PEDERSEN AND
K.F. JØRGENSEN / Introduction 1

Original Abstracts 9

PART 1. THE PHILOSOPHY AND RECENT HISTORY OF
PROBABILITY THEORY AND STATISTICS

N.H. BINGHAM / Probability and Statistics: Some Thoughts at
the Turn of the Millennium 15

    1. Introduction 15
    2. Early History 15
    3. The Rise of Measure Theory: Prelude to the
        Grundbegriffe 16
    4. The Triumph of Measure Theory:
        The Impact of the *Grundbegriffe* 17
    5. The General Theory of Processes
        and Stochastic Integration 19
    6. The Evolution of Statistics in the Twentieth Century:
        Classical Theories 21
    7. The Evolution of Statistics in the Twentieth Century:
        Other Approaches 24
    8. The Development of Applied Probability in the Twentieth
        Century 27
    9. The Impact of Physics 31
    10. The Impact of the Computer 34
    11. The Impact of Finance 36
    12. Kolmogorov's Later Work and Algorithmic
        Information Theory 39
    13. Critique of the Shafer-Vovk Approach via
        Game Theory 41

14. Postscript 43
References 44

VOLODYA VOVK / Kolmogorov's Complexity Conception of Probability 51

1. Introduction 51
2. Kolmogorov's Frequency Interpretation 53
3. Kolmogorov Complexity as Tool for the Frequency Interpretation 56
4. Kolmogorov's Complexity Conception 58
5. Algorithmic Theory of Randomness 61
6. Conclusion 64
Notes 67
References 67

EBERHARD KNOBLOCH / Emile Borel's View of Probability Theory 71

1. Probability Theory and Borel's Philosophy of Mathematics 72
2. Foundations of Probability Theory and Objections Against It 76
3. Borel, von Mises, and Keynes 81
4. Borel and Reichenbach 87
5. Scientific Determinism versus Probabilistic Indeterminism 90
References 93

BERNA EDEN KILINÇ / The Reception of John Venn's Philosophy of Probability 97

1. Venn's Analysis of the Reference Class Problem 98
2. Venn's Reception 105
3. Reference Classes of the Classical Probabilists 111
4. Conclusion 114
Notes 116
Archival Sources 118
References 118

PART 2. CONTEMPORARY ISSUES IN PROBABILITY THEORY AND STATISTICS

J.B. PARIS / On the Distribution of Probability Functions in the Natural World 125

    1. Introduction 125
    2. Univariate Case....................................... 127
    3. The Univariate Model 130
    4. Properties of J 133
    5. Egon Pearson's Investigations 135
    6. The Multivariate Model............................. 137
    7. Conclusions.. 142
    Acknowledgements 143
    Postscript 143
    Notes.. 143
    References... 144

GLENN SHAFER / Nature's Possibilities and Expectations 147

    1. Dynamic Regularity in Nature 148
    2. Nature as an Idealization 151
    3. Towards an Intellectual History of Nature as Ideal
        Witness 154
    4. The Inadequacy of Stochastic Processes.............. 156
    5. A Framework for Causal Debate 157
    6. Determinism and Free Will Within Nature's
        Event Tree 161
    Notes.. 163
    References... 164

T. SEIDENFELD / Remarks on the Theory of Conditional Probability: Some Issues of Finite versus Countable Additivity 167

    1. Introduction 167
    2. Conditional Probability $P(\cdot|A)$ when $P(A) = 0$ 170
    3. Some Finitely Additive Conditional Probability 174

Notes 177
References 177

Index 179

PREFACE

This volume in the *Synthese Library Series* is the result of a conference held at the Roskilde University, Denmark, September 16-18, 1998. The purpose of this meeting was to shed light on some of the recent issues in probability theory and track their history; to analyze their philosophical and mathematical significance, and to analyze the role of mathematical probability theory in other sciences. Hence the conference was called *Probability Theory— Philosophy, Recent History and Relations to Science.*

The editors would like to thank the invited speakers including in alphabetical order Prof. N.H. Bingham (Brunel University), Prof. Berna Kılınç (Bogazici University), Prof. Eberhard Knoblock (Techniche Universität Berlin), Prof. J.B. Paris (University of Manchester), Prof. T. Seidenfeld (Carnegie Mellon University), Prof. Glenn Shafer (Rutgers University) and Prof. Volodya Vovk (University of London) for contributing, in the most lucid and encouraging way, to the fulfillment of the conference aim. The editors are also grateful to the invited speakers for making their contributions available for publication.

The conference was organized by the *Danish Network on the History and Philosophy of Mathematics*

http://mmf.ruc.dk/mathnet/

The editors would like to thank the network's organizing committee consisting of Prof. Kirsti Andersen (University of Aarhus), Prof. Jesper Lützen (University of Copenhagen), Dr. Tinne Hoff Kjeldsen (Roskilde University) and the committee's secretaries Lise Mariane Jeppesen and Jesper Thrane (Roskilde University). In turn, the network was only made possible by a research grant from the *Danish Natural Science Research Council* for which the editors and the network remain very grateful.

We would also like to thank Kluwer Academic Publishers and especially acquisition editor Rudolf Rijgersberg and Jolanda Voogd for their help during the process of publication of this volume.

PREFACE

Finally, the editors are indebted to Prof. Jørgen Larsen from Roskilde University who once more helped us during the typesetting process in $\text{\LaTeX}\,2_\varepsilon$.

Vincent F. Hendricks   Stig Andur Pedersen

Klaus Frovin Jørgensen

Roskilde
February 2001

CONTRIBUTING AUTHORS

In order of appearance:

N.H. Bingham was born in 1945, and studied mathematics at Oxford. He worked for his doctorate in Cambridge under D. G. Kendall; his thesis (1969) was on limit theorems and semigroups in probability theory. He worked for thirty years at the University of London (at Westfield, Royal Holloway and Birkbeck Colleges), serving as Professor first of Mathematics, then of Statistics, before joining Brunel University as Professor of Statistics in 2000. His interests remain focused on probability theory, particularly limit theorems, but he has also worked extensively on analysis (Tauberian theory and regular variation), statistics (non-parametrics), mathematical finance, and history of mathematics.

Volodya Vovk was born and raised in a small town in the Western Ukraine. In 1983 he graduated from Moscow State University (with major in mathematical logic). From 1996 he has been at the Department of Computer Science, Royal Holloway, University of London. His research interests include algorithmic information theory (in particular, algorithmic randomness and predictive complexity), machine learning and foundations of probability.

Eberhard Knobloch, born in 1943, studied mathematics, classical philology, and history of science and technology, since 1980 professor of history of science and technology at the Technical University of Berlin, member of several Academies of Sciences, Honorary Professor at the Northwest University of Xian, PR of China. Research interests: History and philosophy of mathematical sciences, Renaissance technology. Recent publications: *G. W. Leibniz, Hauptschriften zur Versicherungs- und Finanzmathematik*, edited together with J.-M. von der Schulenburg, Berlin 2000 (686 pp.); "Analogy and the Growth of Mathematical Knowledge", in E. Grosholz, H. Breger (eds.), *The Growth of Mathematical Knowledge*, Amsterdam 2000, 295-314.

Berna Kılınç received her B.S. and M.A. in mathematics from Indiana University, Bloomington. In 1997 she completed her

Ph.D. in the Conceptual Foundations of Science Program at the University of Chicago. Her dissertation, entitled *The One and the Many of Frequentism*, is a philosophical and historical examination of the frequency accounts of probability in the nineteenth century. She has published on this topic in *Studies in History and Philosophy of Science* and in *British Journal for the History of Philosophy*. She worked as a post-doctoral fellow in the Max-Planck-Institut für Wissenschaftsgeschichte in Berlin, and is currently employed in the Philosophy Department of Bogazici University, Istanbul, Turkey.

J.B. Paris is Professor of Mathematics at the University of Manchester and the author of *The Uncertain Reasoner's Companion*. He has made contributions to many areas in logic and the foundations of mathematics, in particular in set theory, arithmetic, and most recently uncertain reasoning.

Glenn Shafer is Professor in the Department of Accounting and Information Systems of the Graduate School of Management of Rutgers University. In 1976 Glenn Shafer published *A Mathematical Theory of Evidence*, which formulated the Dempster-Shafer theory, now widely used for expressing judgments of uncertainty in expert systems. In recent years, Glenn Shafer has been particularly interested in expert systems in audit judgment. In 1996, he published a second major book, *The Art of Causal Conjecture*, which is concerned both with expert systems based on causal models and with the empirical investigation of causality. Glenn Shafer is also interested in the history and philosophy of probability and statistics and in the theory of finance. Currently he and a colleague, Volodya Vovk, have a contract with Wiley to publish a book entitled *Probability and Finance: Its Only a Game*. Glenn Shafer has published in journals in statistics, philosophy, history, psychology, computer science, economics, engineering, accounting, and law. He has won teaching awards in both mathematics and business.

T. Seidenfeld, H. A. Simon Professor of Philosophy and Statistics, Departments of Philosophy and Statistics, Carnegie Mellon University. T. Seidenfeld works at the interface between philosophy and statistics, often being concerned with problems that involve multiple decision makers. For example, in collaboration with M.J. Schervish and J.B. Kadane (Statistics, CMU), they have relaxed the norms of Bayesian theory to permit a unified standard, both for individuals acting as separate decision makers and collectively, in forming a cooperative "group" agent. By contrast, this is an impossibility for strict Bayesian theory. For a second example, in collaboration with Larry Wasserman (Statistics, CMU), they have examined the short-run consequences of using Bayes' rule for updating a set of "expert" Bayesian opinions with shared information. They focus on anomalous cases (they call "dilation"), where an experiment is certain to result in new evidence that increases the experts' uncertainty about an event of common interest—where "uncertainty" is reflected in the extent of probabilistic disagreements among the experts. His current collaboration with Kadane and Schervish includes a theory for indexing the "degree" of incoherence in a non-Bayesian statistical decisions.

V.F. HENDRICKS, S.A. PEDERSEN AND K.F. JØRGENSEN

INTRODUCTION

Since the measure theoretical definition of probability was put forward by Kolmogorov, probability theory has developed into a mature mathematical theory. Today it is a fruitful mathematical field with many important applications in philosophy, science and engineering to mention but a few fields of application. In spite of this, many mathematicians, statisticians, and philosophers still feel uneasy with the measure theoretical definition of probability. Even Kolmogorov himself felt so dissatisfied with his definition that he initiated a new mathematical approach to probability based on the concept of complexity. Furthermore, many scientists and philosophers have realized that there exist several rather different concepts of probability, all covering a spectrum from subjective belief to objective frequencies. And although the axiomatic view covers important properties of all these concepts, there still exist important features of probability which have not been captured by the measure theoretical axioms. These problems regarding a comprehensive understanding of the probability concept have, in the twentieth century, led to several interesting theories. Just to mention a few of these: von Mises' and Wald's work on the so-called collectives, probability concepts in quantum theory, subjective probability, Bayesian concepts, complexity based concepts, and the relationship between logic and probability.

The title *Probability Theory—Philosophy, Recent History and Relations to Science* is a reflective title of the contributions to be found in this volume. The volume is accordingly divided into two parts: The first part entitled *The Philosophy and Recent History of Probability Theory and Statistics* including articles by N.H. Bingham, Volodya Vovk, Eberhard Knobloch and Berna Eden Kılınç describe and analyze various of the above mentioned aspects of the development of probability theory from the first formulations. The second part *Contemporary Issues in Probability Theory and Statistics* including the articles by Glenn Shafer, J.B.

Paris and T. Seidenfeld respectively take on different contemporary issues like Nature's role in probability theory and trade-offs between finitely additive and countably additive probability measures.

N.H. Bingham's contribution to the volume entitled "Probability Theory and Statistics: Some Thoughts at the Turn of the Millennium" gives a thorough overview of the development of measure theory into probability from the French school of Borel, Fréchet and Lévy, the Polish school of notably Steinhaus, Zygmund and Banach to the Russian school of Khinchin and particularly Kolmogorov with his now classical work *Grundbegriffe der Wahrscheinlichkeitsrechnung* from 1933. The *Grundbegriffe* had a massive impact and in order to thoroughly appreciate this Bingham in detail describes the evolution of statistics and probability theory in the twentieth century with initial emphasis on the classical theories in parametric statistics of F.Y. Edgeworth and R.A. Fisher. Then in non-parametric statistics, functional analysis combined with measure theory give rise to the theory of weak convergence of probability measures which quite painlessly and naturally can be adapted to various empirical distributions and processes. This in turn gives rise to a number of results among others the Glivenko-Cantelli theorem, which by and large is law of large numbers appropriate for non-parametric statistics and also forms the basis for the theory of Glivenko-Cantelli classes. Other important results along these lines include the Kolmogorov-Smirnov theorem and the Erdös-Kack-Donsker invariance principle. To further emphasize the role of measure theory, Bingham continues by describing various other approaches to statistics in the twentieth century including most notably Bayesian and decision theoretic approaches and how the two theories can be integrated. Today probability theory enjoys a wide range of applications; but also the other way around: Statistics and probability theory have been greatly influenced by a variety of other fields; by queing theory, mathematical biology and genetics, survival analysis, reliability theory; also by physics where probability theory has profited from results in statistical mechanics, quantum theory, rates and turbulence studies; Bingham then also places applied

INTRODUCTION

probability in the context of computability theory like in Monte Carlo methods, computer-intensive methods in statistics and randomized algorithms; finally, Bingham describes the vast areas in finance and economics which have contributed to the development of probability theory and statistics. The article is concluded by a critique of Shafer and Vovk's conception of probability theory based on a game-theoretic approach rather than on the measure theoretic idea discussed by Bingham.

Bingham's overview is followed by Volodya Vovk's contribution "Kolmogorov's Complexity Conception of Probability" which initially discusses the distinction between Kolmogorov's theory of probability on the one hand as a branch of measure theory and the foundations for its application on the other which according to Kolmogorov should be based on frequentism. Kolmogorov essentially believed that the mere theory of probability was exhausted in the axioms stated in the *Grundbegriffe* together with a few additional conditions. With respect to the foundational issue of the application of probability theory Kolmogorov later changed the motivation for his frequentistic approach when he realized that computability theory, the theory of algorithms and in particular the algorithmic theory of randomness provide a foundation for the frequentistic approach. These insights combined led Kolmogorov to his complexity conception of probability especially through the discovery of universal Turing machines. Vovk thoroughly analyzes Kolmogorov's complexity conception of probability and describes its development. Then Vovk points to various advantages but also shortcommings of the complexity approach and finally discusses some general issues pertaining to the modern algorithmic theory of randomness.

The historical part continues with two contributions also devoted to more particular historical and philosophical topics— Eberhard Knobloch's discussion on Emile Borel's view of probability theory and Berna Eden Kılınç's analysis of the lack of enthusiasm pertaining to the reception of Venn's philosophy of probability.

Even though Emile Borel is mostly known as one of the founders of measure theory itself he actually also published particularly and quite extensively on probability theory between 1905-1950. The foundations of Borel's probability theory is the subject of Eberhard Knobloch's contribution "Emile Borel's View of Probability Theory". Despite Borel's extensive achievements in mathematics his motivation for probability theory is his view of probability theory as a kind of social and empirical science rather than as an intrinsically mathematical discipline. Through a series of examples ranging from the method of majorities over linguistic problems to social morality Knobloch demonstrates these social and empirical motivational grounds for Borel's conception of probability. In addition Borel stressed on numerous occasions that the fundamental notion of probability is largely subjective and rests on our willingness to accept bets on single cases. Given this assumption, Borel obviously runs counter to the celebrated views of R. von Mises, J.M. Keynes and H. Reichenbach who all thought that statements concerning the probability of single cases are either senseless in themselves, presuppose repetition to make sense or true by being posits rather than assertions as Reichenbach suggested. Yet through another series of systematic examples, Knobloch shows how Borel thoroughly disagress with all these suggestions. Finally, Knobloch discusses how Borel could reconcile the goal of science which is prediction concerning the supposedly determinate laws of nature with the indeterminism of his own view and probability theory in general.

In his treatise from 1866, *The Logic of Chance*, John Venn formulated the first systematic account of the frequency approach to probabilities. It was received then and even looked upon later with some apprehension. Berna Eden Kılınç analyzes the reasons for this first apprehensive receception in her contribution "The Reception of John Venn's Philosophy of Probability". In the paper Kılınç argues that the main reason for not accepting the frequentistic Vennian approach to probabilities is to be found in his treatment of the *reference class problem*. More specifically, many felt that even though Venn was the first to analyze the problem in philosophical depth, the Vennian suggestions for a solution fixed by *natural kinds* was inadequate and unsatisfactory due to the rather vague prescriptions for characterizing the populations required for reliable statistical inference that the natural kinds provide. Kılınç demonstrates that both philosophers and logicians sympathetic to Venn's frequentism, like C.S. Peirce and critics, including F.H. Bradley, F.Y. Edgeworth and J.M. Keynes largely subscribe to that same criticism. From here Kılınç compares Venn's frequentism to the classical views of probabilities, in particular A. Quetelet's position, and the way in which the classical probabilists treat the reference class problem. In the end Kılınç concludes that converting actual statistical practice into a frequentistic theory of probability was complicated by ontological problems and most notably by the introduction of large populations of entities with similar properties through universals like natural kinds.

This concludes the philosophical and historical part of the volume. The latter part devoted to contemporary issues centers particularly around two themes: One of the important questions in contemporary probability theory is the relation between Nature and the probabilities which are assigned to the events she might eventually show us. The classical question is of course whether probabilities exist in Nature, but the contemporary issue is rather whether is it possible to model chance, objective probability, causality and even our own probabilistic reasoning by assuming that Nature in a way distributes probabilities over various possible events for us to uncover or make predictions about. One approach to this issue is discussed by J.B. Paris in his article "On the Distribution of Probability Functions in the Natural World." The aim of the contribution is exactly to model the distribution of what Paris calls "natural" probability functions, that is functions that inquirers actually find in the real world and which are then used to draw statistical inferences. Paris' idea is to model complicated 0-1 random processes in which the randomness is hidden very deep by large sentences of a propositional logic. Based on this model it is possible to construct a "natural" distribution of expected values of the complicated 0-1 random processes. Besides this natural modelling, the approach is also interesting because it sheds new light on the relation between logic and probability theory, the classical Bayesian prior problem and the general issue of reasoning under uncertainty. Besides this natural modelling, the approach is also interesting because it sheds new light on the relation between logic and probability theory, the classical Bayesian prior problem and the general issue of reasoning under uncertainty.

Yet another aspect of the relation between Nature and probability is discussed by Glenn Shafer in his contribution "Nature's Possibilities and Expectations." The article describes and philosophically justifies an approach to probabilistic reasoning in which Nature is an active actor rather than a passive by-stander one is trying to approach through probabilistic reasoning. From this perspective, Nature witnesses the real unfolding of events and can thus make predictions and has expectations which never fail. As inquiring agents we hardly ever see the events as Nature does but nevertheless often make conjectures about Nature's predictions which may be depicted as nodes in Nature's "event tree". Nature thus becomes an ideal witness and predictor and by our ongoing approximations to her event tree, Nature is the idealized limit of actual and potential inquiring agents. In turn, claims about what Nature can or cannot predict become claims about our own achievements, ambitions and capacities to predict in the scientific process. The idea of event trees also provide a flexible framework for representing causal structures since it conveniently leaves us free to decide whether Nature witnesses a causal regularity or not. This is demonstrated through a number of illustrative examples. Shafer concludes by discussing the pertinent issue of determinism vs. indeterminism as seen in the light of Nature's event tree and how this issue relates to the concepts of causality and objective probability.

In the final paper of this volume, "Remarks on the theory of conditional probability: Some issues of finite versus countable additivity", T. Seidenfeld discusses some important trade-offs between a finitely additive probability measure and a countably additive probability measure. Countable additivity is a natural and required assumption for modern measure theory. On the other hand finite additivity seems to suffice for actual probability problems while countable additivity may lead to paradoxes. Kolmogorov himself entertained the idea that probability should be countably additive, or infinitely additive as he labeled it, but at the same called this very assumption a "useful" tool. Conversely, de Finetti and Savage have argued that probability is only mandated to being finitely additive. Seidenfeld initially lists and simultaneously discusses a number of reasons for considering the theory of probability only finitely additive. Then by means of comparison it is analyzed how theories of conditional probability based respectively on a finitely additive and a countably additive probability measure fair with respect to various principles of probability in particular the so-called Principle of $\aleph_0$-Conglomerability.

ORIGINAL ABSTRACTS

Even though some of the final written contributions diverge in terms of content from the original conference presentations we have decided to include the original abstracts to complete the conference proceedings.

N.H. Bingham, *Measure into Probability: From Lebesgue to Kolmogorov*

Both probability and statistics had emerged as subjects by the nineteenth century, but neither had achieved maturity or its modern form. This talk traces the evolution of probability, and statistics, towards our current conception of them, with particular reference to the impact of measure theory, from its introduction in definitive form by Lebesgue in 1902 to its successful use in the axiomatisation of probability by Kolmogorov in his *Grundbegriffe* of 1933.

We begin with the situation in the late nineteenth century and around the turn of this century. We continue with the French school (Borel, Fréchet, Lévy, ...), the Polish (Steinhaus, Zygmund, Marcinkiewicz, ...), the Russian (Bernstein, Khinchin, Kolmogorov, ...), Italian (Cantelli, de Finetti, ...), German (Hausdorff, von Mises, ...), American (Wiener, ...), English (Keynes, Fisher, Ramsey, Jeffreys, ...), and finally Kolmogorov's classic *Grundbegriffe der Wahrscheinlichkeitsrechnung* of 1933.

Volodya Vovk, *Kolmogorov's Complexity Conception of Probability*

Kolmogorov's goal in proposing his complexity conception of probability was to provide a better foundation for the *applications* of probability (as opposed to the *theory* of probability; he believed that his 1933 axioms were enough for the theory of probability). The complexity conception was a natural development of Kolmogorov's earlier frequentist conception combined with (a) Turing's discovery of the universal computing device, and (b) realization that only finite data sequences are of any interest in the applications of probability. Besides the complexity conception itself, I will discuss its elaborations by Martin-Löf, Levin *et al* and will make an attempt to analyze their advantages and limitations.

Eberhard Knobloch, *Emile Borel's View of Probability Theory*

Apart from many textbooks Borel published more than fifty papers between 1905 and 1950 on the calculus of probability. He stressed its applications to the different sociological, biological, physical, and mathematical sciences. The applications are the true realities. Hence, his conception of probability theory depended on this realistic conception of mathematics. While he was mainly motivated by Poincaré and Bertrand, he rejected the frequency of interpretations of probability by von Mises and Reichenbach and criticized Keynes' logical approach, because Keynes did not say anything about the applications of the calculus of probability to the physical sciences. The paper will deal with Borel's notion of probability, compared to these contemporary other conceptions, with the implications of his philosophy of mathematics for probability theory, and with his theory of determinisms on different scales.

Berna Eden Kılınç, *Why was John Venn's Work on Probability Almost Forgotten in our Century?*

John Venn's Logic of Chance, though well known and respected in Britain, did not have a wide following in the nineteenth or twentieth centuries. Venn was well acquainted with the new generation of logicians and statisticians in Britain, Francis Herbert Bratley, Francis Y. Edgeworth and John Maynard Keynes, but the latter did not subscribe to the theory Venn had developed. Surveying their arguments, I argue that the reference class problem was the main reason why Venn's theory did not dominate the philosophy of probability at the turn of the century. The frequency view of probability, glossing over periodic and non-periodic change, idealizing away temporality, was not the most congenial measure of uncertainty, especially for the theoreticians of social sciences. Confronted with the problem of historical change which dissolved timeless average men, Venn's only solution was to consider ideal series. Classical probabilists, however, were better equipped to deal with historicity, and it was their way of measuring uncertainties which informed Venn's critics. This reception story indicates a gap between the rise of probabilities. Some recent works on the history of statistical reasoning, for example, Theodore Porter's *The Rise of Statistical Thinking*, and Ian Hacking's *The Taming of Chance*, describe a rapid and massive spread of statistical practice in the nineteenth century. At the backdrop of this expansion of statistical thinking, my history of frequentism is not a triumph story. At the end of the nineteenth century, frequentism was only one of the several acceptable accounts of probability. My account does not conflict with that of Porter or Hacking, but accentuates the difficulties of converting statistical practice into a frequency theory of probability.

J.B. Paris, *On the Distribution of Probability Functions in the Natural World*

The purpose of my talk will be to describe the underlying insights and results obtained by myself and my colleagues, in a series of papers aimed at addressing the 'Prior Problem' by *modelling* the distribution of 'natural' probability functions, that is those probability functions on $\{0,1\}^n$ which we encounter naturally in the real world. I shall compare the consequences of the model we obtained with, in particular, another widely adopted model, the Dirichlet distribution(s), and indicate that it appears in some ways superior.

Glenn Shafer, *The Unity and Diversity of Probability*

The basic story of probability, the story about a game between Statistician and Reality, can be used in many different ways. One way leads to subjective probability, another to objective or causal probability. The notion of causal probability depends of the idea of an ideal scientist whose probabilities play out in practice. This talk will trace the historical antecedents of this idea and will contrast it historically and philosophically with its chief rival, the paradoxical idea of counterfactual reality.

T. Seidenfeld, *Should Probability be Countably Additive? Some Trade-offs in the Theory of Conditional Probability*

This presentation examines two problem areas for the philosophy of probability:

- Though appearing to be little more than a formal matter about infinity, the question whether probability should be countably additive (as in Kolmogorov's theory) or mandated only to be finitely additive (as in the theories of de Finetti and Savage) is shown to have significant consequences for several basic principles of induction and decision.
- A generalization of Bayesian theory, the theory of Robust Bayesian methods using sets of probabilities, is examined for several of its surprising consequences. For one example, the familiar Bayesian result that cost-free data have non-negative expected utility for simple decisions is shown not to obtain in Robust Bayesian theory.

PART 1

THE PHILOSOPHY AND RECENT HISTORY OF PROBABILITY THEORY AND STATISTICS

N.H. BINGHAM

PROBABILITY AND STATISTICS: SOME THOUGHTS AT THE TURN OF THE MILLENNIUM

1. INTRODUCTION

This paper forms my written contribution to the proceedings of the conference on *Probability Theory—Philosophy, Recent History and Relations to Sciences*, held at Roskilde University from 16-18 September 1999 and ably organised by Professor Stig Andur Pedersen. I spoke at the conference on the title 'Measure into Probability: From Lebesgue to Kolmogorov'. The written version of this talk has now appeared (Bingham, 2000a) in the series *Studies in the History of Probability and Statistics*, published by the journal *Biometrika*, as number XLVI (Studies XLVI below). The present paper is written with three purposes in mind:

1. to complement Studies XLVI, particularly by covering the rest of the twentieth century;
2. in doing so to give a view, however partial or personal, of the state of the subject at the turn of the millennium;
3. to contribute to the discussion of the novel game-theoretic approach presented at the conference by Glenn Shafer and Volodya Vovk (see Vovk's contribution to this volume).

2. EARLY HISTORY

The view we take of probability and statistics is to regard them as the two sides of the same coin: the mathematics of randomness, or uncertainty, and of what information we can glean when our available data is random/uncertain/obtained by sampling/subject to measurement error, etc. The field is vast and protean. To obtain any overall view, it is – as usual – best to stand back first, and look at the historical roots of the field and its subsequent evolution.

Those interested in the history of probability and statistics are fortunate in having to hand a number of excellent books, and the papers of a number of distinguished scholars. The earliest book on the subject, Todhunter (1865), is now an historical artefact in its own right; for a variety of modern scholarly views, see e.g. Stigler (1986), Porter (1986), Daston (1988), Hald (1990, 1998), Johnson and Kotz (1997). To the further writings of these authors, one might add the works of Ian Hacking and of O.B. Sheynin. For the twentieth century, our main focus here, one has in addition a flood of biographical and autobiographical information, together with the primary textbook and journal literature.

3. THE RISE OF MEASURE THEORY: PRELUDE TO THE *GRUNDBEGRIFFE*

Although the story it outlines is standard and well-known, we summarise here in broad outline the viewpoint and content of Studies XLVI. Whether one accepts it (as I do) or not (like Shafer and Vovk), it is not in contention that the standard approach to probability is measure-theoretic. The unsatisfactory nature of probability theory at the beginning of the twentieth century is aptly commemorated in Problem 6 of Hilbert's famous list of problems given to the International Congress of Mathematicians in Paris in 1900. The emergence soon thereafter of the mathematical machinery of measure theory, in the hands of Borel, Lebesgue and others, laid the basis for the measure-theoretic approach to probability. Important early work was done in 1909 by Borel himself - most notably, Borel's form of the strong law of large numbers, and his Normal Number Theorem: almost all numbers are strongly normal to all bases simultaneously (thus all ten digits $0, 1, 2, .., 9$ appear with equal asymptotic frequency in their decimal expansion, and similarly for the two digits $0, 1$ in their dyadic expansion; likewise for duodecimal expansions, etc.). The exceptional set – of (Lebesgue) measure zero – is clearly needed here, to cover such obvious exceptions as the rationals. This is a highly non-constructive existence theorem: what is remarkable from a mathematical point of view is that to this day, no specific example of such a number is known. What is remarkable from an historical point of view is that Borel, one of the founding fathers of measure theory, largely avoided use of measure theory in the proofs of his results.

The penetration of measure theory into probability theory had to await the work of a number of authors in several different mathematical schools. From the French school came the work of

Fréchet, one of the pioneers of the abstract approach to analysis, and later of Lévy, who wrote the first textbook account of probability from the measure-theoretic standpoint in 1925. In Poland, Steinhaus gave in 1923 an early attempt at an axiomatization of probability using the language of measure theory. His work was continued by such eminent Polish colleagues as Marcinkiewicz, Zygmund, Banach and Kac. In Russia, the work of Markov early in the century (Markov chains date from 1908) was continued by Bernstein, Khinchin and Kolmogorov, Khinchin's law of the iterated logarithm of 1924 being a particularly striking contribution.

Several other countries contributed during this period: Italian workers such as Cantelli and de Finetti, Germans such as Hausdorffand von Mises, Americans such as Wiener, Feller (a Yugoslav, and refugee from the Nazi period) and Doob. English contributions were principally on the statistics side, most importantly the great work of R. A. Fisher from the 1920s on.

The turning-point in this long incubation period was the publication in 1933 of Kolmogorov's classic book *Grundbegriffe der Wahrscheinlichkeitsrechnung*, (Kolmogorov, 1933) (translated into English as *Foundations of the Theory of Probability* in 1955). Here Kolmogorov sets up and develops the basic framework standard of today: the use of a probability space – triple of sample space, sigma-field of events and probability measure – as a mathematical model of a random experiment; the interpretation of probability as a measure of mass one, expectation as integral, random variable as measurable function, almost-sure event as set of full measure, etc. For elaboration of this rather skeletal account, and references, we refer to Studies XLVI.

4. THE TRIUMPH OF MEASURE THEORY: THE IMPACT OF THE *GRUNDBEGRIFFE*

It is as well to take stock of the mathematical and probabilistic environment at this time.

On the mathematical front, the most relevant area was analysis. The new measure theory and measure-theoretic (Lebesgue) integral had been available for a generation, but had not yet been adopted everywhere; meanwhile, the new field of functional analysis was in rapid development following the publication in 1932 of Banach's classic book *Théorie des Opérations Linéaires*. A key step in convincing a still sceptical and conservative mathematical public that integration had to go measure-theoretic was taken by Norbert Wiener (1894-1964), with the publication in 1933 of his book *The Fourier Integral and Certain of its Applications*. This book, together with Wiener's work of around this time on Tauberian theorems and on generalised harmonic analysis – with its applications to time series in statistics – showed beyond doubt that the Fourier integral was an essential weapon in the mathematician's armoury, and that its successful use demanded the Lebesgue integral.

On the probabilistic front, some were quick to realise the importance of the measure-theoretic approach. An early example is Harald Cramér (1893-1985), whose Cambridge Tract of 1937, *Random Variables and Probability Distributions*, was a highly influential textbook exposition of the theory seen at that time from the viewpoint of the Kolmogorov axiomatics. After the War, Cramér followed this with his 1946 book *Mathematical Methods of Statistics*, the first successful textbook synthesis of the new ideas of Kolmogorov on the probability side and of Fisher – likelihood, maximum likelihood, sufficiency, etc., on the statistics side. From the historical point of view, an interesting human account of this period is contained in the autobiographical article Cramér (1976).

The basic framework of the Kolmogorov axiomatics is used to handle probability theory in a static setting. In a dynamic setting, where randomness unfolds with time, one needs more, and this is the task of the theory of stochastic processes. The first systematic exposition of stochastic process theory in the new context following the *Grundbegriffe* was given by the American probabilist J.L. (Joe) Doob (1910-) in his classic book of 1953, *Stochastic processes*. Doob's introduction contains the flat, bald statement "Probability is simply a branch of measure theory, with its own special emphasis and field of application, and no attempt has been made to sugar-coat this fact". Doob's book, most noted for its famous chapter on martingales, is still in use – at least for reference

– nearly half a century later, a highly unusual feat in a subject as dynamic as mathematics. The quotation above is good evidence of the thoroughness with which the post-*Grundbegriffe* measure-theoretic approach permeated the subject in the two decades after its publication.

Further light is thrown on the development of probability theory during this period by the classic two-volume book by William Feller (1906-1970), *An Introduction to Probability Theory and its Applications*. Volume one, in its editions of 1950, 1957 and 1968, avoids the need for measure theory by restricting itself to the discrete setting, where probabilities can be evaluated by summation. Despite the restrictions of this setting, the book remains a goldmine of interesting examples to this day, and was deeply influential on a generation of probabilists, including the present writer (1945-). The second volume, in its 1966 and 1971 editions, has also been highly influential—compared to almost anything except the first volume. It handles continuous probability, but stops short of any full discussion of the dynamic framework needed to handle stochastic processes. Feller had intended a third volume on stochastic processes, but died before the 1971 edition of the second volume was finished.

In parallel with the appearance of modern textbooks, the literature of this period included such leading journals as the *Annals of Mathematical Statistics* (1930-72), before its bifurcation into the *Annals of Probability* and the *Annals of Statistics* in 1973.

5. THE GENERAL THEORY OF PROCESSES AND STOCHASTIC INTEGRATION

The phenomenon of Brownian motion was observed under the microscope by the Scottish botanist Robert Brown (1773-1858) in 1828, though he had been anticipated by Leeuwenhoek and others. The subject attracted attention from various applied viewpoints (to which we return below), but found its definitive theoretical formulation at the hands of Wiener in the years 1920-23. Regarded from the point of view of the theory of stochastic processes, Brownian motion is accordingly called the Wiener process in his honour; let us denote it by $W = (W(t))$.

The idea of forming a *stochastic integral* – an integral in which a path $W(t)$ of the Wiener process is used as an *integrator* – can be traced to the work of Wiener in 1933 and Doob in 1937 (see e.g. Bingham (1998) for references and details). But this concept is generally associated with the name of the Japanese probabilist Kiyosi Itô (1915-), who introduced his Itô-integral in 1944 (Itô's work, which allows random non-anticipating integrands, is more general than that of Wiener or Doob). The resulting Itô calculus differs strikingly from ordinary (Newton-Leibniz) calculus, by the presence of extra Itô-terms, whose roles reflect the quadratic variation property of Brownian paths.

The sigma-field generated by a random variable X is a construct found in the *Grundbegriffe*, and it has long been known from Doob's work that this sigma-field carries the interpretation of 'the information contained in X'. Consequently, in the dynamic framework of a stochastic process $X = (X(t))$, where the randomness unfolds with time, one envisages an increasing family of sigma-fields, one for each time t, carrying the interpretation of 'the information contained in all random variables seen by time t'. Such a family is called a *filtration*. The term, and the concept, is due to Paul-André Meyer, the founder of the Strasbourg school of probability and the *general theory of processes*. A filtration – subject to suitable conditions (the 'usual conditions' or 'conditions habituelles' of Meyer) – is now recognised as a necessary ingredient, together with a probability space, of a setting in which a stochastic process can be suitably defined. One speaks accordingly of a *filtered probability space*.

This setting, together with the classification of stopping times (it would be out of place to elaborate on this here) and of sigma-fields, constitutes the skeleton of the general theory of processes, developed by Meyer (1966) in his book *Probability and Potentials* of 1966, subsequently elaborated in the five-volume book of the same title written with Claude Dellacherie. The resulting machinery has proved to be what is needed to develop a theory of stochastic integration, and stochastic differential equations, flexible enough to handle the multiplicity of demands imposed by the need for a coherent theory and the demands of the many applications. A magisterial synthesis was provided in Meyer (1976). Itô calculus has now emerged, stripped down to its bare essentials, as one of the most potent tools of modern probability theory. There are many excellent textbook accounts; we refer in particular to that of Rogers and Williams (1987).

The trilogy of Williams (1991) (measure-theoretic probability and martingales), Rogers and Williams (1994) (stochastic processes—foundations) and Rogers and Williams (1987) (stochastic processes—Itô calculus) provides not only an excellent place to learn and source for reference, but a good example, from the historical viewpoint, of the state of the subject at the turn of the millennium.

6. THE EVOLUTION OF STATISTICS IN THE TWENTIETH CENTURY: CLASSICAL THEORIES

We refer to Stigler (1986) for a fine account of the development of statistics up to the end of the nineteenth century. Much had been accomplished, highlights being the distribution theory of the normal (Gaussian, Laplacian) distribution, in any number of dimensions - the theory of the multivariate normal distribution is largely the achievement of F.Y. Edgeworth (1845-1926)—and the role of the linear model, including the key ideas of regression and correlation—largely the achievement of Francis Galton (1822-1911) and Karl Pearson (1857-1936).

It is perhaps surprising, looking back, that such things as a clear distinction between parameter values and the corresponding sample quantities had to await the work of R.A. Fisher (1890-1962), in his foundational works from 1922 on. Fisher's great work—on likelihood and sufficiency, analysis of variance, design of experiments, maximum likelihood estimation and the role of information, and much else, revolutionised the field. It is difficult to overstate either Fisher's impact on statistics or his genius. As well as the depth and importance of his work, the breadth of his interests is simply staggering: quite apart from his work on statistics, Fisher made profound contributions to genetics. He was, with J.B.S. Haldane and S. Wright, one of the three founding fathers of modern genetics. For background, see e.g. the biography J.F.Box (1978). For appreciations of Fisher and his work in statistics, see e.g. Savage (1976) and the references cited there, Seidenfeld (1979).

Statistics as we have discussed it so far is parametric statistics, in which one seeks to describe the population under study, or the source of randomness or uncertainty, in terms of a finite—preferably, a small—number of parameters. The prototypes are the normal distributions—describable in one dimension via two parameters, the mean and the variance, and in higher dimensions via a mean vector and a covariance matrix. For many purposes, such parametric models are adequate to describe the populations under study, and are simple enough to be tractable, since the mathematics involved is finite-dimensional, and highly developed.

The problem is that *all models are wrong*—parametric models in particular. For example: it would be idle to deny the usefulness and importance of the normal distributions. Nevertheless, no real population is exactly normal—and even if it were, sampling from it would not yield an exactly normal sample, if only for the trite reason that writing down the values sampled would necessarily involve truncation—rounding to the number of significant figures being used; this would yield only rational values, whereas the normal distribution, merely because it possesses a density, yields only irrational values when sampled, with probability one.

One is thus presented with a dilemma. One either has to embrace a parametric model—even if one does not believe in it—for the sake of convenience, and hope that the results of using it do not depend too sensitively on the details of the model assumed—raising questions of *robustness*—or one has to do without a (parametric) model, in which case everything immediately becomes infinite-dimensional. Naturally, the mathematics of infinite-dimensional situations is much more complicated than that of finite-dimensional ones.

Fortunately, the mathematical machinery needed to handle non-parametric statistics exists. Functional analysis—which we mentioned earlier in connection with the work of Banach—combined with measure theory, gives us, for example, the theory of *weak convergence* of probability measures. This is well adapted to the study of *empirical* distributions and processes. Recall that the empirical distribution of a random sample of size n is the (random) probability distribution with mass $1/n$ on each sample point. The Fundamental Theorem of Statistics (Glivenko-Cantelli Theorem: V. Glivenko (1897-1940) and F.P. Cantelli (1875-1966), both in 1933) tells us that as the sample size n increases, the empirical distribution converges to the actual – population – distribution F, uniformly on the whole line , with probability one. To summarise loosely: *The sample determines the population in the limit.* In higher dimensions, matters are more complicated, because of the richer geometry. Suffice it to say that this Glivenko-Cantelli theorem – the 'law of large numbers' appropriate for the setting of non-parametric statistics – gives rise to the theory of Glivenko-Cantelli classes—classes of sets for which such a theorem applies (prototype: the half-lines in one dimension). Corresponding to this law of large numbers, there is a central limit theorem—the Kolmogorov-Smirnov theorem, which gives the basis for *tests of goodness of fit* in this setting. This result finds its natural setting in the context of weak convergence theory for probability measures, and in particular for the Erdös-Kac-Donsker *invariance principle*. In higher dimensions, one has a corresponding theory of Donsker classes. The associated theory of weak convergence illustrates beautifully the symbiotic relationship between probability and statistics. It has been well expounded in a series of excellent and influential texts; see e.g. Billingsley (1968), Shorack and Wellner (1986), van der Vaart and Wellner (1996)and Dudley (1999).

Sometimes it is appropriate to compromise between these two approaches, and use a semi-parametric framework, in which the model has both a parametric and a non-parametric part. For background, see e.g. Bickel et al. (1993).

7. THE EVOLUTION OF STATISTICS IN THE TWENTIETH CENTURY: OTHER APPROACHES

7.1. Bayesian Statistics

In classical statistics as described above, one visualises oneself as sampling from some unknown population, and seeking to gather information about this population as one samples from it. Here, the source of the randomness is in the *act of sampling*.

Natural though this mental framework is, it is by no means inevitable. One can invert the entire picture. Suppose that one chooses to speak, not in terms of randomness, but of *uncertainty*. In the context above, the data in the sample are regarded as random—indeed, if one re-sampled, or if another statistician did a similar sampling operation, different data would be obtained. However, for one statistician who has drawn one sample, the data, now obtained and to hand, are better regarded as *known* than as random. The data constitute one of the two things the statistician knows. The other is the knowledge, or belief, that he brings with him when he sets out to study the population. No one invests the time and trouble to undertake a serious statistical study in total ignorance. One brings with one some degree of understanding—often a very shrewd one. But one's knowledge is incomplete—or one would not need to undertake the study. One thinks in terms of one's uncertainty—the *prior uncertainty*, prior because it reflects one's knowledge *before* sampling. To a Bayesian statistician, the proper way to describe uncertainty, or uncertain knowledge, is in terms of a *probability distribution*. Thus a Bayesian statistician begins his analysis by choosing a *prior distribution* to describe his beliefs, or knowledge, or uncertainty, before sampling. He then draws a sample, just as a classical statistician above would do so. Finally, he uses Bayes' Theorem (the Reverend Thomas Bayes (1702-61), posthumously in 1763) to *update* his beliefs/knowledge/uncertainty to give his *posterior distribution*, representing his position *after* sampling. For a detailed historical account, see Dale (1999); for an excellent textbook account of modern Bayesian statistics, see Robert (1994). Of course, a statement of Bayes' Theorem – mathematically, a two-line consequence of the definition of conditional probability – is contained in any elementary text on probability or statistics; it can be summarised in the easily memorisable form

Posterior density $\propto$ prior density times likelihood.

One thus sees that the essence of the Bayesian approach, or the Bayesian *paradigm*, as its adherents like to call it, is the use of Bayes' Theorem to update one's view systematically, as new information becomes available. The parallel with ordinary life is irresistible: One wakes up in the morning, as the 'me' of a particular day. One goes forth into the world, gathering new information in the course of the day's work. One sleeps, assimilating this new experience (one dreams, while doing this). One wakes up again, as the updated 'me' of the next day—and so on.

Personal note. I have to confess that the parallel between Bayesian updating and the updating of ordinary life is for me not only extremely attractive but all but convincing. I nevertheless decline to describe myself as a Bayesian, partly because one is then expected to conform to the tenets of the creed. For example, I like to talk of unbiased estimates at will. A Bayesian does not allow himself to do this. Unbiasedness is not a Bayesian concept. It involves an expectation, over all data, potential as well as actual; Bayesian concepts restrict themselves to the two components of prior belief and data actually obtained.

One further aspect worth mentioning here is that the vast majority of Bayesian statistics relates to parametrics rather than to non-parametrics. This is beginning to change, but for the present, a statistician such as myself whose primary interests lie on the non-parametric side is likely to remain at most a 'crypto-Bayesian'.

The Likelihood Principle. This doctrine that all that one should be influenced by – apart from one's prior beliefs – are the data one actually obtains, and in particular the likelihood function determined by one's data, is known as the Likelihood Principle. It was first enunciated clearly by George A. Barnard, in papers from 1947 to 1949 ; later work has been done in this area by A. Birnbaum and others. For a monograph account, see

Berger and Wolpert (1988) (including historical background in section 3.2).

7.2. Decision Theory

It is always sound, when contemplating a task, to think carefully about what one will do with the outcome. Much of (classical) parametric statistics falls under one of the two broad headings of estimation of parameters and testing of hypotheses. A decision theorist will attempt to quantify the adverse consequences of incorrectly estimating a parameter value, or incorrectly rejecting/failing to reject some hypothesis, by appropriate choice of a *loss function* (or, to look at things the other way about, a *utility function*). He will then proceed by some decision rule. Depending on the statistician, he may seek to minimise his expected loss, minimise his maximum loss ('minimax'), etc. For a textbook account of the decision-theoretic approach to statistics, see e.g. the classic book by Ferguson (1967).

The decision-theoretic and Bayesian approaches can be integrated; this is the viewpoint developed in Robert (1994). See also the 1985 book *Taking Decisions* (Lindley, 1985) by D.V. Lindley (1923-), one of the fathers of Bayesian statistics in Britain. The essence of this link is the idea of *minimising expected loss* (or maximising expected utility), the heart of the theory of *coherent* decision-making, which developed from the work of F.P. Ramsey (1904-1930) (published posthumously in Ramsey (1931)). The minimax approach to decision theory also has affinities with game theory. This subject is originally the creation of J. von Neumann (1903-57); see the classic von Neumann and Morgenstern (1947). For example, in the theory of the two-person zero-sum game, the Fundamental Theorem of Game Theory tells us that the game has a value; there is a corresponding equilibrium, where the two opponents each play optimally and achieve a minimax. The interplay between game theory and decision theory is developed in, e.g. Blackwell and Girshick (1954), Luce and Raiffa (1957). One should note, however, that game theory and Bayesian statistics do not combine well. This is because, to quote Lindley, 'The Bayesian paradigm is severely limited to a single decision maker.

One of the world's most difficult and important problems is to extend the Bayesian argument to decision-makers in conflict: to have an axiom system for conflict (Smith, 1995, 318).

As mentioned in the Introduction, a new approach to probability via game theory has recently been developed by Shafer and Vovk, and is the subject of Vovk's contribution to this volume. We return to this aspect below. Suffice it to say for now that the idea of visualising a game between the statistician and reality has deep classical roots.

A propos of minimax theory, Lindley comments that he has never understood how such ideas, whose game-theoretic origin is so clear, have become so widely adopted in statistics. One might adapt this remark to say that the Shafer-Vovk game-theoretic approach—to which we return in section 12—sits uneasily with the Bayesian paradigm.

7.3. Incompletely Specified Probabilities

Both the classical and Bayesian paradigms in statistics, in their different ways, need the randomness or uncertainty to be described by a probability distribution. The question arises as to whether a theory can usefully be developed in which the mechanism generating the randomness, or uncertainty, may be, not just unknown, but not fully specified at all. Theories of this kind have been expounded in the books of Walley (1991) and Paris (1994).

8. THE DEVELOPMENT OF APPLIED PROBABILITY IN THE TWENTIETH CENTURY

We live in a complicated and uncertain world, which constantly presents us with challenging problems as to how to proceed in the face of this uncertainty—or randomness. Probability theory provides us with the theoretical tools. But, perhaps more than with any other branch of mathematics, it is the applications – the demands of describing the real world—that drive the theory. Applied probability has developed into a vast field; we describe some of the main lines briefly below. But in general terms one should say that the use of stochastic modelling to describe and study the world around us in all its unpredictability and complexity is one of the great themes of modern applied mathematics. The reader wishing to augment the necessarily skeletal remarks below could do no better than to consult the pages of the *Journal of Applied Probability* (1964-) and its sister journal, the *Advances in Applied Probability* (1969-), as well as *Annals of Applied Probability* (1991-).

1. Queuing Theory. The emergence of the telephone provided one of the principal early stimuli towards the development of queuing theory. The Danish telephone engineer Agner Krarup Erlang (1878-1929), who published from 1909 on, was perhaps the first person whose work is still cited in the queuing literature; for an account of his life and work, see Brockmeyer et al. (1948). Questions of congestion naturally arose in the context of the early telephone because all calls needed to go through the operator. Under the economic and social pressures of ordinary life, organisers of facilities where customers, patients etc. were liable to have to queue became aware of the need to harness theory to practical needs in this area. One of the early workers here was D. G. Kendall (1918-), who established the notation still used in the subject (M/G/1, etc.) in his survey of 1951. This was inspired by the needs of the Berlin Airlift of 1948-49, where the ability of the Allies to supply West Berlin, then blockaded by Stalin, was limited by the available runways, rather than aircraft, aircrew or supplies. The subject of queuing theory has grown enormously since then, and many authors have made important contributions; in addition to Kendall and Lindley, we may mention here L. Takács and J.W. Cohen. An excellent general account of applied probability, with particular reference to queuing theory, is Asmussen (1987). For links with stochastic process theory in general and martingale theory in particular, see Brémaud (1981), Baccelli and Brémaud (1994). It is interesting to note that the stochastic calculus used here is different from the Itô calculus based on the Wiener process. The natural tool here is instead the Palm calculus based on the Poisson process (which, although easier, developed later). Palm calculus (the name derives from work of C. Palm in 1943 on intensity fluctuations in telephone traffic) was developed in direct response to the needs of queuing, and is a good example of the ability of applications to stimulate theory. The theory of one queue with a single server is now well developed; the theory of the many-server queue is much harder and more fragmentary. However, much of the emphasis in more recent work is on networks of queues. Obvious applications include telephone networks, more recent telecommunication networks such as the Internet and the World-Wide Web, highway networks etc. For a recent survey, see e.g. Williams (1998).

2. Point Processes: Earthquakes, Volcanoes, ... Such events as the occurrence of a telephone call, a fire, accident or insurance claim, divide up the time axis by the arrival of random points. The aspect of stochastic process theory most relevant here is that of point processes. The excellent monograph Daley and Vere-Jones (1988) opens with an interesting historical account, including in particular the work of Palm mentioned above. The theory has important geophysical applications: both the authors of the book cited were motivated by the prediction and other problems concerning events such as earthquakes and volcanoes. Again, apart from the human and historical interest, the point to emphasise is the symbiotic relationship between theory and applications so characteristic of probability.

3. Mathematical Biology and Mathematical Genetics. The two great themes of nineteenth-century biology were the Darwinian theory of natural selection and the Mendelian theory of genetics. This last was originally overlooked, then rediscovered around the turn of the twentieth century. It was at first believed that the Darwinian and Mendelian positions were incompatible. Eventually, however, the modern view, which synthesises the two, emerged. The three central figures in the founding of modern mathematical genetics were the British scientists J.B.S. Haldane (1892-1964) and R.A. (Sir Ronald) Fisher (1890-1962), and the American Sewall Wright. One of the earliest mathematical contributions to this area was the Hardy-Weinburg law of 1908, something of an oddity as G.,' Hardy (1877-1947) was a pure mathematician— analyst and number theorist—notorious for his lack of interest in applications. Following the development of Markov chains, it became clear that this theory was just what is needed for the problems of mathematical genetics. This theme appears already in the book of Feller (1968), and is developed at length in, e.g., Ewens (1979). In addition, branching processes, stochastic population models, stochastic models for competition and predation, species abundance, epidemics, and many other topics, have been studied with great success by probabilistic methods. Excellent textbook accounts are to be found in, e.g., Bartlett (1960) and Bailey (1964). It is worth noting that the demands of mathematical genetics led to the development of a whole new branch of pure mathematics: non-associative algebras known as genetic algebras. These, together with many of the topics mentioned above, were central to the interests of Philip Holgate (1937-93); for background and details, see e.g. the obituary article Bingham (2000b).

4. Survival Analysis. Imagine a medical study devoted to the treatment of a potentially lethal disease. Patients enter a clinical trial and are given appropriate treatments, which are recorded; the progress of patients is monitored through time. Meanwhile, the cohort of patients under treatment is winnowed, partly by withdrawal (through moving away, declining treatment because of problems with side-effects, etc.), partly by death. The data available to a statistician advising the medical team is thus of a special kind: it is censored. This censoring is characteristic of data encountered in this area of statistics, which is known as survival analysis. Important early work was done in this area by the English statistician D.R. (Sir David) Cox, with his proportional hazard's model of 1972. It became realised, through the work of O. Aalen and others, that martingale methods were perfectly adapted to coping with the technical problems of censoring. The area has been intensively studied, because of both its practical importance in medicine and its mathematical interest. For a full account, with references, see the monograph Andersen et al. (1993). Again, the thing to emphasise is the symbiotic relationship between theory and applications, and the wide and valuable applicability of probability.

5. Reliability Theory. Why do we age? Why do machines have limited useful lifetimes? To address such questions, one needs to view a biological organism, or a machine, as a complex assemblage of components, some essential and some not, some duplicated and some not, linked in complicated ways—series, parallel, etc. Each component has its own lifetime distribution (the theory of lifetime distributions generates its own vocabulary—of survivor functions, hazard rates etc.). Qualitative properties of lifetime distributions may be identified; in reliability theory, they receive acronymic names such as NBU (new better than used), IFRA (increasing failure rate on average), etc. The theory addresses the question of how features of the lifetime distributions of components, and of the interconnections between them, translate into features of the lifetime distribution of the complete organism, machine, etc. The subject has been well developed; see e.g. the books of Barlow and Proschan (1965, 1975). It uses an interesting blend of probability and statistics to address questions of fundamental interest in both the biological and the engineering sciences. Rather different questions arise in other aspects of reliability theory. The consequences of catastrophic breakdown in such things as nuclear power plants, jumbo jets, supertankers, bridges, dams, sea defences etc. are so serious that great efforts are worthwhile to try to quantify, and minimise, the dangers involved. Because serious accidents are so rare, the available data are sparse. It is the task of the probabilist/statistician to extract the greatest possible value from the very limited data available in such cases. This is the province of Extreme Value Theory, by now a well-developed field. For background and applications, see e.g. Embrechts et al. (1997) and the references cited there.

9. THE IMPACT OF PHYSICS

1. Statistical Mechanics. The statistical view within physics was pioneered in the nineteenth century by James Clerk Maxwell (1831-1879), Josiah Willard Gibbs (1839-1903) and Ludwig Boltzmann (1844-1906), in their work on thermodynamics and gas theory. In the twentieth century, Albert Einstein (1879-1955) used

Brownian motion in physics in 1905, leading to the Einstein relation: the mean square displacement of a diffusing particle grows in proportion to time, the constant of proportionality—the diffusion coefficient—being informative about Avogadro's number. Einstein's work, together with that of Smoluchowski, Langevin and Perrin, was instrumental in establishing convincingly that matter consists of atoms and molecules - indeed, that heat and temperature relate to the energy with which molecules of a fluid or gas vibrate and collide with each other at random. For a fine account of Einstein's work, and its significance, see Pais (1982). As a result of the thermal oscillations above, one observes spontaneous fluctuations of physical quantities. These may be handled by using Brownian motion as the driving noise process. See e.g. (Fowler, 1936, XX), Bingham (1998). More recent aspects of statistical mechanics, motivated by such phenomena as ferromagnetism, have become increasingly important in probability theory. See for instance the books of Ellis (1985) on large-deviation theory and Georgii (1988) on Gibbs' states. Of great interest to both physicists and mathematicians are questions concerning phase changes. Particularly relevant here is the theory of percolation, for which see e.g. Grimmett (1989).

2. Quantum Theory. According to the Copenhagen interpretation of quantum mechanics, the squared modulus of the wave function bears the interpretation of a probability density; see e.g. (Landau and Lifshitz, 1977, III). Thus the only statements one can make in quantum physics are statistical ones. Indeed, in quantum statistical mechanics, a statistical approach is even more forced, since particles no longer possess an individual identity (Landau and Lifshitz, 1977, IX). Thus, quantum theory – the most important scientific achievement of the twentieth century – necessitates a viewpoint which is inherently statistical—or probabilistic, if one prefers. In passing: it is well-known that Einstein was unable to come to terms with quantum theory, because, as he put it, he refused to believe that 'God plays dice with the universe' (Pais, 1982, VI). To a probabilist, such as myself, nothing could be more natural than to picture God playing dice with the universe: What else would He do with it? An alternative approach to quantum theory was developed by R.P. Feynman (1918-1988), in terms of what he called 'sums over histories', or path-integrals. Mathematically, these are functional integrals with respect to Wiener measure. The functional integral approach to quantum theory has been developed in the books of Simon (1979) and Glimm and Jaffe (1987):

Quantization = Integration over function space.

(Glimm and Jaffe, 1987, Ch. 8).

3. Rates. The theory of stochastic processes in general, and of Poisson processes in particular, gives us a tool for modelling events which occur quite unpredictably—'completely at random', or 'out of the blue'. One builds a Poisson process (in time and/or space, possibly inhomogeneous) in terms of *rates*—propensities of things that can happen to happen. Such rates may have physical interpretations—for example, in rates of radioactive decay, or of quantum-mechanical tunnelling. Typically, such rates measure how exponentially small probabilities of exponentially unlikely events are. The theoretical context is that of *large deviation theory*; for a textbook account (motivated by the physics of ferromagnetism), see e.g. Ellis (1985).

4. Turbulence. Whether in air, where it affects air traffic, or in water, turbulence is a phenomenon of prime importance. It is known to be an extremely complex one. Very influential early work was done by Kolmogorov in the 1940s; for this, and later work of G. I. Taylor, see e.g. Batchelor (1990, 1996). Kolmogorov's basic insight was to visualise turbulence as a process of a localized excess of energy cascading downwards from larger to smaller scales of length. Interestingly, this idea has a parallel in the study of market behaviour, where it is information, rather than energy, that cascades downwards, as the impact of price-sensitive information is felt from global newsflash down to company level and below (O. E. Barndorff-Nielsen, personal communication).

10. THE IMPACT OF THE COMPUTER

1. Monte Carlo. Frequently in probability and statistics one encounters problems where one cannot obtain explicit analytic solutions. In such circumstances, one may simulate instead: conduct a computer experiment. This is the so-called Monte Carlo method, developed around 1944 in Los Alamos by Ulam, von Neumann and others; see e.g. Ulam (1976) for historical background, Hammersley and Handscomb (1964), Morgan (1984) for textbook accounts. In order to simulate, one needs an inexhaustible supply of computer-generated random numbers. In fact, of course, such numbers cannot be other than deterministic, strictly speaking, but by use of suitable number-theoretic devices such as congruential generators, 'pseudo-random numbers' —numbers produced deterministically, but with appropriate statistical properties to pass muster as random in practice—can be produced as needed. See e.g. Niederreiter (1992) for a full account and references. The Monte Carlo method and simulation have become ubiquitous in recent decades, particularly with the explosion in computing power. This alone, in my opinion, is enough to caution one against 'waxing too theological' about what is random and what isn't. When two situations cannot be told apart, but one is known to be deterministic while the other is obtained by sampling, it may well be sensible to treat them both as random and interchangeable, depending on context and the purpose one has in mind.

Aside: π. The combination of sophisticated mathematics—due to Gauss, Ramanujan and others—and modern supercomputers has allowed computation of the decimal expansion of the constant $\pi = 3.14159\ldots$ to literally billions of digits. (To a number theorist, the only natural way to expand π is as a continued fraction, so expanding it decimally is a sterile exercise, but let that pass.) Nothing could be more deterministic, or less random, than π—expanded or otherwise. Nevertheless, this vast sequence of expansion coefficients provides ample raw material for subjection to the whole gamut of statistical tests for randomness available - which it passes with flying colours. As the Chudnovsky brothers (among the pioneers of expansion of π) put it, 'π is a damned good fake of a random number'. This is a good illustration of a recurring theme: what actually is random may be a matter of principle, but how things behave, or should be treated, is also a matter of practicalities and context. (My date of birth is 19.03.1945; the expansion of π, started after 19,031,945 places, is quite determistic—indeed, predictable—to me or to anyone who knows this, but looks quite random to anyone else.).

2. *Markov Chain Monte Carlo.* It frequently happens that one wishes to sample from a distribution which is too complicated to obtain explicitly. One possible approach is to seek to represent this distribution as the limiting, or ergodic, distribution of a Markov chain, and sample instead from the Markov chain. In many cases, such a procedure is quite tractable. The method is known as 'Markov chain Monte Carlo' or 'MCMC'. Of course, MCMC generates its own problems, largely concerned with how long one needs to run the Markov chain for before one can safely regard it as having reached equilibrium. The MCMC methodology combines very naturally with that of Bayesian statistics. The resulting synthesis includes such methods as the 'Gibbs sampler' (the idea is suggested by statistical mechanics; the statistical impetus was in image processing). The method finds ever-widening fields of application. Medical statistics is particularly important here: one can handle large numbers of dimensions (which may correspond to patients), and complicated dependence relationships—e.g., for a multi-factorial disease—which can be handled by graphical methods (see e.g. Lauritzen (1996)). The power of the method can only be judged by its applications; for a range of case studies, see Gilks et al. (1996).

3. *Computer-Intensive Methods in Statistics.* The ready availability of computer power has stimulated statistical methods specifically designed to exploit it, to do things that were previously beyond reach. One such area is that of re-sampling schemes such as the bootstrap, jackknife etc. The method was introduced by Efron (1979); for a textbook synthesis, see Hall (1992). Another

area where computer power has revolutionised practice is in applications of the Fast Fourier Transform (FFT). This has greatly increased the practical usefulness of Fourier-based methods such as frequency domain methods in time series analysis. For an interesting application in queuing theory, see Grübel (1991). Another example again is provided by wavelets. Traditionally in time series, one had to choose between time-domain methods, and frequency-domain methods based on Fourier analysis. No more: wavelets provide an alternative, or complement, to Fourier methods which enables the statistician to handle time and frequency aspects together. They also provide vast scope for data compression—to the extent that the FBI now uses wavelets to compress and store its data bank of criminals' fingerprints. (What makes this possible is that a typical function can be well described by only a few wavelet coefficients. By contrast, sparseness of Fourier coefficients corresponds to pathological behaviour of the function.) For background and references, see e.g. Carmona et al. (1998).

4. Randomized Algorithms. One of the major mathematical developments of recent years has been the growing realization that many tasks which are perfectly deterministic can nevertheless be undertaken more efficiently by probabilistic methods than by deterministic ones. One example is in testing large numbers for primality: it is easier to generate 'probabilistic primes' – numbers that pass enough primality tests to be prime in the overwhelming majority of cases – than genuine primes. This area has important applications in cryptography. For an excellent treatment of the general area of randomized algorithms, see Motwani and Raghavan (1995).

11. THE IMPACT OF FINANCE

Broadly speaking, much of economics is concerned with how prices are determined, while in finance prices are taken as given. It is only in the last half-century or so that mathematical methods have made an impact on finance. The first major contribution was that of Harry Markowitz in 1952. Markowitz's thesis was, again broadly speaking, that a financial agent—investor, portfolio manager etc.—needs to think in terms of both *return* and *risk*, and that the way to reduce risk is to hold a diversified portfolio, some assets in which are negatively correlated with others. Now return is basically measured in terms of *means*, while risk is measured in terms of *variances*; thus the Markowitz approach is known as *mean-variance analysis*. From this time on, the methods, language, viewpoint and results of probability have penetrated the field of finance, and irrevocably changed it.

Markowitz's approach necessitated the calculation of mean vectors and covariance matrices for whatever basket of assets the investor was contemplating. Since the range of assets available as candidates for investment is huge, this presents formidable numerical challenges even nowadays in the computer age; when Markowitz was writing, before the widespread use of computers, this was a crucial drawback. One way to reduce this computational burden is to compare each asset, not with all other assets but with some common benchmark or index, serving as a proxy for the market as a whole. The idea of working in terms of the correlation of each asset with 'the market' is known as the Capital Asset Pricing Model (CAPM, or 'cap-emm'), and was introduced by Sharpe in 1964, Lintner in 1965 and Mossin in 1966. The impact of these ideas is sometimes known as 'the first revolution in mathematical finance'.

The 'second revolution in mathematical finance' stems from the work of Black & Scholes, and of Merton, both in 1973, on pricing of options and other financial instruments such as warrants. Such instruments are known as financial *derivatives*, as they are derived from underlying fundamentals such as the prices of stock, of bonds, etc. The prototypical result in this field is the *Black-Scholes formula*, which gives, in explicit closed form, the rational price for an option on a stock, in terms of things one has to hand such as the stock price, the interest rate and the present time, and also one parameter, known as the *volatility*, which serves to describe how sensitive the stock price is to new information. This formula was first derived by formulating and solving the Black-Scholes partial differential equation; Merton showed how PDE methods could be avoided, in place of methods based on arbitrage arguments and dynamic hedging - continually rebalancing one's portfolio appropriately.

Arbitrage is basically the exploitation of opportunities for making risk-free profit. Naturally, those who trade at prices that expose them to such attack by arbitrageurs are liable to be eliminated from the market—and hence, such arbitrage opportunities as may arise are liable to be dissipated by the market activity of those moving to exploit them. Arguments based on such ideas (which may perhaps seem trite at first sight) are powerful. They go back at least as far as the Modigliani-Miller theorem of the early 1950s. In the hands of S.A. Ross and others in the late 1970s, they developed into the Arbitrage Pricing Technique (APT). By 1981, arbitrage ideas had been linked with the mathematics of martingales and stochastic integration, in the work of J.M. Harrison, D. Krebs, S.R. Pliska and others, who identified the concept of *equivalent martingale measures*. In brief: to price derivatives, one first discounts everything (long standard actuarial practice), and then *takes expectations under the equivalent martingale measure*. This technique is known as *risk-neutral valuation*; for a recent textbook account, see e.g. Bingham and Kiesel (1998).

Thirty years ago, pricing and hedging *any* derivatives was an unsolved problem. Nowadays, pricing and hedging standard (so-called *vanilla*) options and the like is routine, and the market constantly develops new, more complicated ones (so-called exotic options). The open-ended escalation of mathematical complexity needed to develop innovative financial products results in an ever-escalating increase in the mathematical input – from probability theory, statistics and numerics – into the financial services industry. The father of modern stochastic-integration theory, Paul-André Meyer, has often remarked that the development of stochastic integration could have been designed with the needs of mathematical finance in mind—but it wasn't. Instead, this interplay provides another excellent example of a phenomenon common in the history of mathematics—how new theory seems to arrive ready-made for the needs of new applications. One thinks of the spectral theory developed by Hilbert and his school, which could have been – but wasn't – developed with the needs of quantum mechanics – specifically, the wave mechanics formulation of Schrödinger – in mind.

The upshot is that the modern financial services industry is heavily dependent on modern mathematics in general, and on probability and statistics in particular. This is a salutary realization, as it shows that much of what once appeared arcane, theoretical and of academic interest only is in fact as practically useful in the real world of trading and financial affairs as one could wish. For those interested in learning more about this remarkable conduit for mathematical theory into financial practice, we recommend the fascinating popular accounts of Bernstein (1992, 1996).

12. KOLMOGOROV'S LATER WORK AND ALGORITHMIC INFORMATION THEORY

As mentioned in the introduction, one of the motivations of this paper is to comment on the new approach to probability due to Glenn Shafer and Volodya Vovk. We begin by discussing its predecessors.

The prehistory of this approach is to be found in the work of the great Russian probabilist Andrei Nikolaevich Kolmogorov (1903-1987), and his precursors, particularly Richard von Mises (1883-1953). In his book *Wahrscheinlichkeit, Statistik und Wahrheit (Probability, Statistics and Truth)* of 1928, von Mises attempted to construct a theory of probability via so-called *collectives* (von Mises (1928)). This was an attempt to construct a theory – of probability in general, of random sequences in particular – on a *sequence-by-sequence basis*, addressing in particular the question: given a specific sequence, is it (or can it be treated as) random? Natural though this question is, the von Mises theory did not succeed in providing a definitive approach. Not only is his theory not rigorous as mathematics, von Mises did not regard probability as fully mathematical (nor even as mathematicisable). His definition of collectives amounts in effect to an attempt to turn the law of large numbers from a theorem into an axiom. The great French probabilist Paul Lévy (1886-1971) remarked that it is as impossible to build a satisfactory theory in this way as it is to square a circle.

Kolmogorov, in his papers from the 1920s on and in his classic book, the *Grundbegriffe* of 1933, harnessed instead the measure-theoretic approach that became and remains standard. This deals with the totality of all possible outcomes - sequences, say, but has nothing to say about a specific individual outcome. Kolmogorov's approach is the subject of the second volume of his *Selected Works, Theory of Probability and Mathematical Statistics* (TPMS: Kolmogorov (1986). Let us call this approach 'Kolmogorov Mark I', or Kolmogorov I.

It is a tribute to the genius and versatility of Kolmogorov that he was dissatisfied enough with his own creation to return to the subject of probability from a quite different standpoint. In the 1960s, he developed a new approach to probability which does give a sequence-by-sequence treatment of random sequences. In brief, this theory identifies randomness of sequences with *maximal complexity*. For instance, consider the classical situation of a thousand replications of the toss of a fair coin. All possible strings of a thousand 0s and 1s carry equal probability, by symmetry, and are on the same footing from the point of view of the classical theory. But the difference between, say, the sequence of a thousand 0s, and one obtained by actual coin-tossing (in a previous experiment, say) is obvious, to the man in the street as much as (or more than!) to the expert. Kolmogorov saw that there was a fundamental difference to be identified, and that it lay in the complexity of description. The first sequence can be described by two bits of information ('all zeros'), the second needs a thousand. The resulting theory of Kolmogorov complexity formed the subject of the third volume of his *Selected Works, Theory of Information and Theory of Algorithms* (TITA - Kolmogorov (1987)). Let us call this theory Kolmogorov (Mark) II. It has led to the development of the field of *algorithmic information theory*. Contributors include members of the Kolmogorov school such as Vovk, and in the West, workers including Gregory J. Chaitin and Jorma Rissanen.

This theory allows one to give sharp answers to questions as to how far from maximal complexity a sequence can fall short but still exhibit the characteristic properties of randomness. See Kolmogorov and Uspensky (1987) for versions of the strong law of large numbers in this setting, and its sequel Vovk (1987) for versions of the law of the iterated logarithm. These results provided recognisable analogues of the classical strong limit theorems of probability theory—the strong law of large numbers and the law of the iterated logarithm—but without their characteristic qualification 'a.s.'—almost surely/with probability one/outside an exceptional set of measure zero.

In passing, the parallel with the philosophical work of Ludwig Wittgenstein – who created a new and highly acclaimed approach to philosophy (Wittgenstein I), and then became dissatisfied with it and created another, incompatible with his first (Wittgenstein II) – is irresistible.

13. CRITIQUE OF THE SHAFER-VOVK APPROACH VIA GAME THEORY

In the Shafer-Vovk approach (SV below for brevity), one visualises a game between the statistician and nature (not in itself innovative, of course). Their approach to the strong law of large numbers may be summarised by saying that the statistician can force an outcome in which *either* the conclusion of the strong law of large numbers holds—convergence of the relevant sample mean to its appropriate expectation—*or* the statistician achieves a cumulative gain in the game which diverges to infinity. As one interprets this latter outcome as not to be observed in practice, one obtains an interpretation of the strong law of large numbers as representing what will actually happen, *and this without the need for an exceptional set of measure zero*. Not only the strong limit theorems mentioned above, but also their weak counterparts—the weak law of large numbers and the central limit theorem—find their place in this theory. There is some discussion of stochastic processes such as diffusions in this framework, using the machinery of non-standard analysis.

Shafer and Vovk apply their theory also to mathematical finance. In particular, they obtain the Black-Scholes partial differential equation.

Any theory that can produce versions of the classic limit theorems of probability theory on the one hand, and of the Black-Scholes theory of mathematical finance on the other, deserves to be taken seriously. The Shafer-Vovk theory shares the directness of approach of the Kolmogorov second theory, and broadens our range of choices in approaching the treatment of probability. It is a pleasure to salute an impressively original approach, for which the authors deserve our thanks.

My reservations – which are considerable – lie not so much with the Shafer-Vovk theory itself, but with the authors' claims for it and against its measure-theoretic predecessor.

1. The authors claim – repeatedly and with some emphasis – that measure-theoretic probability is complicated, unnatural, cumbersome, and even ugly. The obvious response is that measure theory is a necessary part of the equipment of a fully-trained professional mathematician, if only because integration concepts less advanced than the measure-theoretic Lebesgue integral, such as the Riemann integral, may be easier to learn but are harder to handle and manipulate. One thinks of the ease with which Lebesgue's theorems of monotone or dominated convergence handle questions of interchange of limit and integral, for example. One thinks also of Wiener's use of the Lebesgue theory of integration to handle Fourier integrals—his work on generalised harmonic analysis (1930) and Tauberian theorems (1932), and his books on Fourier integrals (Wiener (1933), Paley and Wiener (1934))—which convinced a still sceptical mathematical public that one had to make the investment of learning measure theory, if only to do integration adequately. This remains as true in the early 2000s as it was in the early 1930s. To anyone knowing measure theory, seeing measure and integral put to use as probability and expectation is natural and painless.

2. The achievements of the measure-theoretic approach to probability – in pure probability, applied probability, statistics, and

indeed pure mathematics – remain as impressive an edifice after the Shafer-Vovk attack as they were before. Being several decades more developed, measure-theoretic probability has been able to tackle a vastly broader range of problems, theoretical and practical, than the Shafer-Vovk game-theoretic approach. This is not said in criticism of the SV theory, for which it is still early days yet. It is said in criticism of the attacks by Shafer and Vovk on the measure-theoretic theory. They say of their own theory 'We can do everything that the measure theorists can do'. The vast range of applications that the measure-theoretic approach has so successfully handled to date – of which it was our pleasant task to discuss a small range above – provides a challenge which many years of work may enable SV to rise to. At the moment, however, the degrees of maturity of the two approaches are so different that a fair comparison is premature; meanwhile, a confrontational challenge of this kind between SV and measure-theoretic probability is rash verging upon suicidal. Such would indeed be typically the case for any new theory facing an established one. In any case, for myself I prefer a more inclusive and less confrontational approach.

14. POSTSCRIPT

The twentieth century proved a period of tremendous growth, for science and mathematics in general and for probability and statistics specifically. In particular, it witnessed the life and work of Kolmogorov and Fisher, respectively the greatest of all probabilists and of all statisticians. Progress has been as dramatic on the applied side, greatly enhanced by the ready availability of modern computing power, as on the theoretical side. The prospects for the twenty-first century are as exciting as they are uncertain. On the probability side, my own prediction is that this young subject – which only came of age in the 1900s – will continue towards achieving the settled maturity that one associates with subjects – such as complex analysis, for example – that came of age much earlier. On the statistics side, the combination of theory and computing power seems sure to produce ever wider and deeper applications. I anticipate further progress by the Bayesian paradigm on the parametric side; the extent to which this may be complemented on the non-parametric side is one of the imponderables that will have to be left to the judgement of posterity.

Department of Mathematical Sciences
Brunel University
England

REFERENCES

Accardi, L. and Heyde, C. (eds) (1998). *Probability towards 2000*, Vol. 128 of *Lecture Notes in Statistics*, Springer.

Andersen, P., Borgan, O., Gill, R. and Keiding, N. (1993). *Statistical Models Based on Counting Processes*, Springer.

Asmussen, S. (1987). *Applied Probability and Queues*, Wiley.

Baccelli, F. and Brémaud, P. (1994). *Elements of Queueing Theory. Palm-Martingale Calculus and Stochastic Recurrences*, Springer.

Bailey, N. (1964). *The Elements of Stochastic Processes, with Applications to the Natural Sciences*, Wiley.

Banach, S. (1932). *Théorie des Opérations Linéaires*, Monografje Matematyczne, Warsaw.

Barlow, R. and Proschan, F. (1965). *Mathematical Theory of Reliability*, Wiley.

Barlow, R. and Proschan, F. (1975). *Statistical Theory of Reliability and Life Testing*, Holt, Rinehart and Winston.

Bartlett, M. (1960). *Stochastic Population Models*, Methuen.

Batchelor, G. (1990). Kolmogorov's work on turbulence, in (Kendall, 1990, 47-52).

Batchelor, G. (1996). *The Life and Legacy of G.I. Taylor*, Cambridge University Press.

Berger, J. and Wolpert, R. (1988). *The Likelihood Principle*, Vol. 9 of *Lecture Notes – Monographs Series*, 2 edn, Institute of Mathematical Statistics, Hayward, CA.

Bernstein, P. (1992). *Capital Ideas. The Improbable Origins of Modern Wall Street*, The Free Press.

Bernstein, P. (1996). *Against the Gods. The Remarkable Story of Risk*, Wiley.

Bickel, P., Klaasen, C., Ritov, Y. and Wellner, J. (1993). *Efficient and Adaptive Estimation for Semi-Parametric Models*, Johns Hopkins University Press. (Reprinted, Springer, 1998).

Billingsley, P. (1968). *Convergence of Probability Measures*, Wiley.

Bingham, N. (1998). Fluctuations, *The Mathematical Scientist* **23**: 63–73.

Bingham, N. (2000a). Measure into probability: from Lebesgue to Kolmogorov, *Studies in the History of Probability and Statistics XLVI., Biometrika* **87**: 145–156.

Bingham, N. (2000b). Obituary: Philip Holgate, *Bull. London Math. Soc.* **32**: 484–492.

Bingham, N. and Kiesel, R. (1998). *Risk-neutral Valuation. Pricing and Hedging of Financial Derivatives*, Springer.

Blackwell, D. and Girshick, M. (1954). *Theory of Games and Statistical Decisions*, Wiley.

Brémaud, P. (1981). *Point Processes and Queues. Martingale Dynamics*, Springer.

Brockmeyer, E., Halstrom, H. and Jensen, A. (1948). *The Life and Works of A.K. Erlang*, Vol. 2 of *Trans. Dan. Acad. Tech. Sci.*

Carmona, R., Hwang, W.-L. and Torresani, B. (1998). *Practical Time-Frequency Analysis. Gabor and Wavelet Transforms with an Implementation in S, Wavelet Analysis and its Applications*, Academic Press.

Cramér, H. (1937). *Random Variables and Probability Distributions*, Vol. 36 of *Cambridge Tracts in Math*, Cambridge University Press. (2 edn 1962, 3 edn 1970).

Cramér, H. (1946). *Mathematical Methods of Statistics*, Princeton University Press.

Cramér, H. (1976). Half a century with probability theory: Some personal reminiscences, *Ann. Probab.* **4**: 509–546.

Dale, A. (1999). *A History of Inverse Probability, from Thomas Bayes to Karl Pearson*, 2 edn, Springer. (1 edn 1991).

Daley, D. and Vere-Jones, D. (1988). *An Introduction to the Theory of Point Processes*, Springer.

Daston, L. (1988). *Classical Probability in the Enlightenment*, Princeton University Press.

Doob, J. (1953). *Stochastic Processes*, Wiley.

Dudley, R. M. (1999). *Uniform central limit theorems*, Cambridge University Press.

Efron, B. (1979). Bootstrap methods: another look at the jackknife, *Annals of Statistics* **7**: 1–26.

Ellis, R. (1985). *Entropy, Large Deviations and Statistical Mechanics*, Vol. 271 of *Grundl. Math. Wiss.*, Springer.

Embrechts, P., Klüppelberg, C. and Mikosch, T. (1997). *Modelling Extremal Events*, Springer.

Ewens, W. J. (1979). *Mathematical Population Genetics*, Vol. 9 of *Biomathematics*, Springer.

Feller, W. (1968). *An Introduction to Probability Theory and its Applications*, Vol. 1, 3 edn, Wiley. (1 edn 1950, 2 edn 1957).

Feller, W. (1971). *An Introduction to Probability Pheory and its Applications*, Vol. 2, 2 edn, Wiley. (1 edn 1966).

Ferguson, T. (1967). *Mathematical Statistics. A Decision-Theoretic Approach*, Academic Press.

Fowler, R. (1936). *Statistical Mechanics. The Theory of the Properties of Matter in Equilibrium*, 2 edn, Cambridge University Press. (1 edn 1929).

Georgii, H.-O. (1988). *Gibbs Measures and Phase Transitions*, Walter de Gruyter.

Gilks, W., Richardson, S. and Spiegelhalter, D. (1996). *Practical Monte-Carlo Markov Chains*, Chapman and Hall.

Glimm, J. and Jaffe, A. (1987). *Quantum Physics. A Functional Integration Point of view*, 2 edn, Springer. (1 edn 1981).

Grimmett, G. (1989). *Percolation*, Springer.

Grübel, R. (1991). Algorithm as265: GI/G/1 via fast Fourier transform, *J. Roy. Statist. Soc.* **C 40**: 355–365.

Hald, A. (1990). *A History of Probability and Statistics and their Applications before 1750*, Wiley.

Hald, A. (1998). *A History of Mathematical Statistics from 1750 to 1930*, Wiley.

Hall, P. (1992). *The Bootstrap and Edgeworth Expansion*, Springer.

Hammersley, J. and Handscomb, D. (1964). *Monte Carlo Methods*, Chapman and Hall.

J.F.Box (1978). *R.A. Fisher: The Life of a Scientist*, Wiley.

Johnson, N. and Kotz, S. (eds) (1997). *Leading Personalities in Statistical Sciences, from the Seventeenth Century to the Present*, Wiley.

Kendall (1990). Obituary, A.N. Kolmogorov (1903-1987), *Bull. London Math. Soc.* **22**: 31–100.

Kolmogorov, A. (1933). *Grundbegriffe der Wahrscheinlichkeitsrechnung*, Springer. (*Foundations of Probability Theory*, Chelsea, 1950).

Kolmogorov, A. (1986). *Theory of Probability and Mathematical Statistics*, Vol. 2 (in Russian) of *Selected Works*, Nauka, Moscow.

Kolmogorov, A. (1987). *Theory of Information and Theory of Algorithms*, Vol. 3 (in Russian) of *Selected Works*, Nauka, Moscow.

Kolmogorov, A. and Uspensky, V. (1987). Algorithms and randomness, *Theory of Probability and Applications* **32**: 389–412.

Landau, L. and Lifshitz, E. (1977). *Quantum Mechanics (Non-Relativistic Theory)*, Vol. 3 of *Course of Theoretical Physics*, Pergamon. 3 edn.

Lauritzen, S. (1996). *Graphical Models*, Oxford University Press.

Lindley, D. (1985). *Making Decisions*, 2 edn, Wiley. (1 edn 1971).

Luce, R. and Raiffa, H. (1957). *Games and Decisions. Introduction and Critical Survey*, Wiley.

Meyer, P.-A. (1966). *Probability and Potentials*, Blaisdell.

Meyer, P.-A. (1976). *Un cours sur les intégrales stochastiques*, *Séminaire de Probabilites X*, Vol. 511 of *Lecture Notes in Mathematics*, Springer. 245–400.

Morgan, B. (1984). *Elements of Simulation*, Chapman and Hall.

Motwani, R. and Raghavan, P. (1995). *Randomized Algorithms*, Cambridge University Press.

Niederreiter, H. (1992). *Random Number Generation and Quasi-Monte Carlo Methods*, SIAM, Philadelphia, PA.

Pais, A. (1982). *Subtle is the Lord: The Science and the Life of Albert Einstein*, Oxford University Press.

Paley, R. and Wiener, N. (1934). *Fourier Transforms in the Complex domain*, Vol. 19 of *American Math. Society Colloquium Publications*, AMS, Providence, RI.

Paris, J. (1994). *The Uncertain Reasoner's Companion – a Mathematical Perspective*, Cambridge University Press.

Porter, T. (1986). *The Rise of Statistical Thinking 1820-1900*, Princeton University Press.

Ramsey, F. (1931). *The Foundations of Mathematics, and other Logical Essays*, Kegan Paul, London. (ed. R.B. Braithwaite, preface by G.E. Moore).

Robert, C. (1994). *The Bayesian Choice. A Decision-Theoretic Motivation*, Springer.

Rogers, L. and Williams, D. (1987). *Diffusions, Markov Processes and Martingales*, Vol. 2: Itô calculus, Wiley.

Rogers, L. and Williams, D. (1994). *Diffusions, Markov Processes and Martingales*, Vol. 1: Foundations (2nd ed.), Wiley, (1st ed., D. Williams, 1979).

Savage, L. (1976). On re-reading R. A. Fisher, *Annals of Statistics* **2**: 441–500.

Seidenfeld, T. (1979). *Philosophical Problems of Statistical Inference: Learning from R. A. Fisher*, Reidel, Dordrecht.

Shorack, G. and Wellner, J. (1986). *Empirical Processes with Applications in Statistics*, Wiley.

Simon, B. (1979). *Functional Integration and Quantum Physics*, Academic Press.

Smith, A. (1995). A conversation with Dennis Lindley, *Statistical Science* **10**: 305–319.

Stigler, S. (1986). *The History of Statistics. The Measurement of Uncertainty before 1900*, Harvard University Press.

Todhunter, I. (1865). *A History of the Mathematical Theory of Probability, from the Time of Pascal to that of Laplace*. (Reprinted, Chelsea, New York, 1949).

Ulam, S. (1976). *Adventures of a Mathematician*, Charles Scribner's Sons.

van der Vaart, A. and Wellner, J. (1996). *Weak Convergence and Empirical Processes*, Springer.

von Mises, R. (1928). *Wahrscheinlichkeit, Statistik und Wahrheit* (3rd German ed. 1951), *Probability, Statistics and Truth* (2nd revised English ed.), George Allen & Unwin, 1957, repr. Dover, 1981.

von Neumann, J. and Morgenstern, O. (1947). *Theory of Games and Economic Behaviour*, 2 edn, Princeton University Press. (1 edn 1944).

Vovk, V. (1987). Law of the iterated logarithm for random Kolmogorov, or chaotic, sequences, *Theory of Probability and Applications* **32**: 413–425.

Walley, P. (1991). *Statistical Reasoning with Imprecise Probabilities*, Chapman and Hall.

Wiener, N. (1933). *The Fourier Integral and Certain of its Applications*, Cambridge University Press.

Williams, D. (1991). *Probability with Martingales*, Cambridge University Press.

Williams, R. (1998). *Some Recent Developments for Queueing Networks*, in (Accardi and Heyde, 1998, 340–356).

VOLODYA VOVK

KOLMOGOROV'S COMPLEXITY CONCEPTION OF
PROBABILITY

ABSTRACT

Kolmogorov's goal in proposing his complexity conception of probability was to provide a better foundation for the *applications* of probability (as opposed to the *theory* of probability; he believed that his 1933 axioms were sufficient for the theory of probability). The complexity conception was a natural development of Kolmogorov's earlier frequentist conception combined with (a) his conviction that only finite data sequences are of any interest in the applications of probability, and (b) Turing's discovery of the universal computing device. Besides the complexity conception itself, its developments by Martin-Löf, Levin *et al* will be briefly discussed; I will also list some advantages and limitations of Kolmogorov's complexity conception and the algorithmic theory of randomness in general.

1. INTRODUCTION

One of the relatively recent breakthroughs in the foundations of probability is the creation of the modern algorithmic theory of randomness; it has become clear that algorithmic notions play a fundamental role in connecting the mathematical theories of probability and statistics with their applications. It can be argued (Levin *et al*) that the notion of randomness itself belongs not to probability or statistics but to computer science. In this paper I will discuss the evolution of ideas which led to the creation of Kolmogorov's complexity conception of probability, the first version of the algorithmic theory of randomness.

My view of this process is represented in Figure 1. The starting point is Kolmogorov's 1933 axiomatic foundation for probability theory (Kolmogorov, 1933). To connect the axioms with reality, Kolmogorov (1933) proposed two conditions, which he called A and B. In Section 2, I discuss Kolmogorov's frequency interpretation of probability, which is based on Condition A. Later Kolmogorov realized that the theory of algorithms provides a perfect foundation for the frequency interpretation (Section 3). It soon

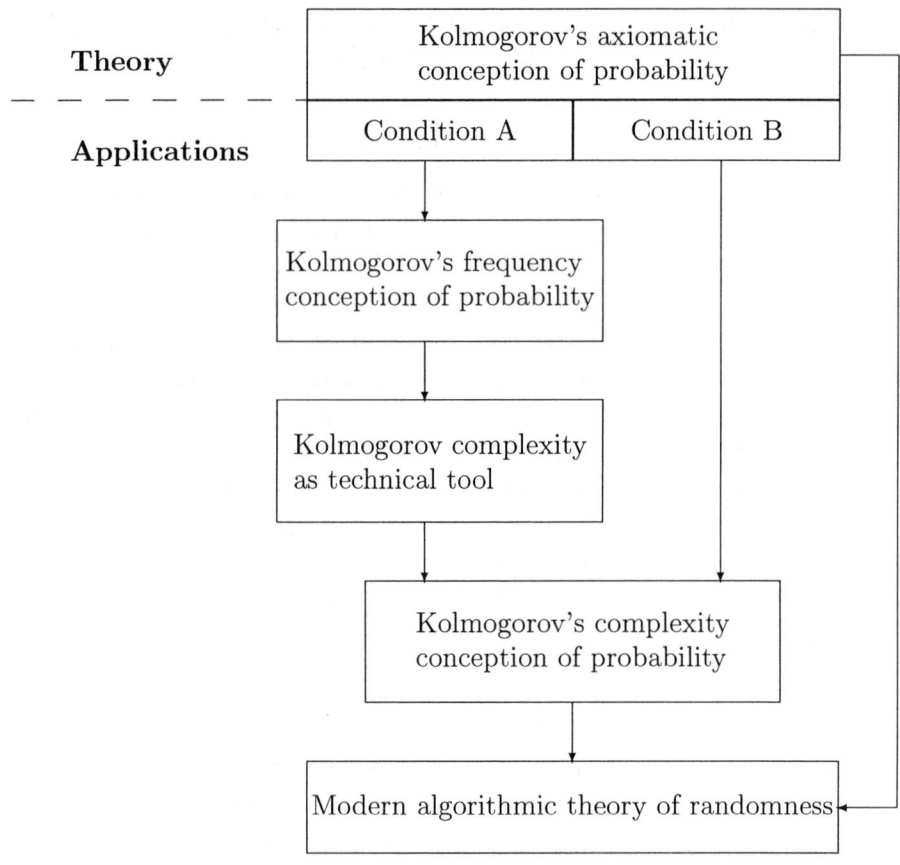

FIGURE 1. Development of the modern algorithmic theory of randomness

turned out, however, that Kolmogorov complexity can be used in a more direct way for connecting the theory and applications, without resorting to frequencies; this is discussed in Section 4. Kolmogorov's complexity conception avoids using the 1933 axioms of probability; predictably, Kolmogorov's students and followers were quick to merge those, which led to the creation of what I call "modern algorithmic theory of randomness" (Section 5).

2. KOLMOGOROV'S FREQUENCY INTERPRETATION

Kolmogorov's original point of view, as expressed in his *Grundbegriffe* (Kolmogorov, 1933), was that

- the theory of probability is just a specialized branch of measure theory,
- the applications of probability should be based on the frequentist conception.

However, any attempt to base probability theory on frequency immediately encounters the usual vicious circle. For example, the frequentist interpretation of an assertion such as $\Pr(E) = 0.6$ is something like: we can be practically (say, 99.9%) sure that in 100,000 trials[1] the relative frequency of success will be within 0.02 of 0.6. But how do we interpret 99.9%? If again using frequency interpretation, we have an infinite regress: Probability is interpreted in terms of frequency and probability, the latter probability is again interpreted in terms of frequency and probability, etc.

Conditions A and B

To avoid the vicious circle, Kolmogorov made a radical step (which, probably, led him a few decades later to his complexity conception of probability), giving a special interpretation to probabilities close to 0 or 1, such as 99.9% above. (For Kolmogorov it was not enough to have a special interpretation of probabilities equal to 0 or 1 precisely, since the latter typically arise only in the case of empirically non-existence infinite sequences.)

The interpretation of probability $\Pr(E)$, where E is an event that can occur in a trial C, offered in Kolmogorov's 1933 book (Kolmogorov, 1933) is given by the following two conditions.

Condition A: One can be practically certain that if C is repeated a large number of times, the relative frequency of E will differ very slightly from $\Pr(E)$.

Condition B: If $\Pr(E)$ is very small, one can be practically certain that when C is carried out only once, the event E will not occur at all.

Condition A stops the infinite regress using the vague "practically certain". Condition B expresses the essence of this idea, stripping it from the unnecessary frequentist context. One might wonder why Kolmogorov needed Condition A at all, since it is implied by Condition B when combined with standard mathematical results (namely, the law of large numbers) and the assumption (implicit in Condition A) of independence of the realizations $C_1, C_2, \ldots$ of the trial C:

$$\left. \begin{array}{l} \text{Condition B} \\ \text{Independence of } C_1, C_2, \ldots \\ \text{Mathematics of probability} \end{array} \right\} \Longrightarrow \text{Condition A}.$$

Probably the answer is that Kolmogorov wanted to give some meaning to stand-alone phrases like "event E will occur with probability 0.6". Condition B, although simpler and more fundamental, does not provide this.

REMARK 1. If we are interested in giving meaning to a probabilistic *theory* (rather than a separate *sentence*), such as quantum mechanics, or the theory "non-deformed European coins land heads up with probability 0.5 and different tosses are independent", or probability predictions issued over time by a probability forecaster, Condition B is sufficient. In the frequentist approach the problem of giving a meaning to sentences is circumvented by replacing the individual probability with a very special theory (the "iid extension" of that individual probability).

Kolmogorov's 1963 Paper

Kolmogorov published several informal expositions of his philosophy of probability in 1938–1959 (Kolmogorov, 1938, 1951, 1956, 1959; Kolmogorov and Gnedenko, 1948), without trying to formalize it. In 1963 he wrote a paper (Kolmogorov, 1963) called "On tables of random numbers", where he did make an attempt to build a formal frequentist theory of probability. He starts that paper by clearly stating the reasons why he had not made such an attempt earlier:

1. The infinitary frequency approach based on *limiting frequency* (as the number of trial goes to infinity) cannot tell

us anything about real applications, where we always deal with finitely many trials.
2. The frequentist approach in the case of large but finite number of trials cannot be developed purely mathematically.

In (Kolmogorov, 1963) Kolmogorov rejected the second statement (but, of course, not the first one, which had always been the basis of his philosophy). Essentially Kolmogorov (1963) developed a finitary version of von Mises' collectives, but instead of "collectives" Kolmogorov was speaking about "(n, ϵ)-random tables".

Consider a "table"

$$T = (t_1 \ldots t_N),$$

which is just a finite binary sequence. Intuitively, T is the sequence of outcomes (for simplicity assumed to be 0 or 1) of independent trials $C_1, C_2, \ldots$. Von Mises's requirements to a (binary) collective were:

1. The limiting frequency of 1s exists.
2. This limiting frequency does not change if we choose an "honest" subsequence.

In the case of finite sequences the first condition is vacuous, and we only need to worry about the second one. In the second condition "does not change" has to be replaced with "does not change much", which leads to the need to speak about "(n, ϵ)-random sequences" rather than just "random sequences". In reality Kolmogorov's definition is even more complicated, since there is an extra parameter $\mathcal{R}_N$ (the set of admissible selection rules, itself a very complicated object), which is not explicitly reflected in his notation. Probably for Kolmogorov a certain lack of mathematical elegance was balanced by the importance of his new philosophy.

As a byproduct, Kolmogorov solved in (Kolmogorov, 1963) some problems with Church's (Church, 1940) explication of von Mises's definition: According to Ville's (Ville, 1939) example, there exist von Mises–Church collectives which violate the law of the iterated logarithm. Remark 2 of (Kolmogorov, 1963) gives an infinitary version of the definition of (n, ϵ)-random sequences; the essential difference from the von Mises–Church definition is that selection rules are allowed to select the elements of the sequence in any order, not necessarily in the direction from left to right. Ville's construction does not work in the case of Kolmogorov's (Kolmogorov, 1963) definition, but there still remains a possibility that even the latter does not avoid similar paradoxes. As shown by Martin-Löf (1966), such possibility is excluded for Kolmogorov's later definitions based on Kolmogorov complexity.

3. KOLMOGOROV COMPLEXITY AS TOOL FOR THE FREQUENCY INTERPRETATION

The "universal" notion of randomness (of course, for finite sequences) was introduced by Kolmogorov in his 1965 paper *Three approaches to the definition of the concept "amount of information"* (Kolmogorov, 1965). The idea is as follows. Let a binary sequence $(t_1 \ldots t_N)$ have k ones. To describe this sequence,

$$\log \binom{N}{k}$$

bits are sufficient, since there at most

$$\binom{N}{k}$$

such sequences. The sequence is *random* if it cannot be described with much fewer bits.

To make this definition precise we need to define the shortest description of a sequence. The key discovery which made this possible was that there is a *universal* method of description, which provides descriptions almost as short as those provided by any alternative method. (This was realized independently and earlier by Ray Solomonoff (1964). The existence of a universal method of description was an implication of the existence of a universal Turing machine.)

The *Kolmogorov complexity* $K(x)$ of x is defined to be the length of the shortest description of x when the universal method of description is used. Allowing the universal method of description to use extra information y, we obtain the definition of conditional Kolmogorov complexity $K(x \mid y)$. Now we can say that a sequence $x = (t_1 \ldots t_N)$ with k ones is *Bernoulli* if $K(x \mid N, k)$ is close to

$$\log \binom{N}{k}.$$

Formally, Kolmogorov defines m-Bernoulli sequences as those satisfying

$$K(x \mid N, k) \geq \log \binom{N}{k} - m.$$

REMARK 2. In (Kolmogorov, 1965, 1968) Kolmogorov used "Bernoulli sequences" in place of "random sequences" used in (Kolmogorov, 1963). From now on we will use "random" to mean "random with respect to the uniform distribution"; when speaking of sequences random with respect to other distributions or classes of distributions, we mention those explicitly.

Kolmogorov was never specific about what to do with his Bernoulli sequences. (One possibility is that he intended to develop a theory similar to the one developed by von Mises for his infinitary collectives.) This makes it difficult to discuss the advantages and limitations of his definition as the foundation for applications of probability. Still it is clear that, whatever his goals were, the definition based on Kolmogorov complexity would serve them better than the von Mises type one given in (Kolmogorov, 1963).

REMARK 3. It should be noticed, however, that even Kolmogorov's later definition does not fully correspond to his motivation (to formalize "coin-tossing sequences", i.e., those that can be obtained as outcomes of fair coin-tossing). It is clear that

Bernoulli sequences $\subseteq$ "coin-tossing sequences".

This inclusion, however, is proper, as the following simple argument shows. When N is large and even and $k = \frac{N}{2}$, the left-hand side of the inclusion contains many sequences, whereas the right-hand side is empty. The details will be given in Section 5.

4. KOLMOGOROV'S COMPLEXITY CONCEPTION

Kolmogorov never abandoned frequentism, but it seems that in his 1983 paper (Kolmogorov, 1983a) (see also the earlier paper (Kolmogorov, 1968) and the paper (Kolmogorov, 1983b)) he came close to it. The paper is called *Combinatorial basis of information theory and probability theory* and was prepared back in 1970 in connection with his talk at the International Mathematical Congress in Nice. In it Kolmogorov gives the most detailed exposition of the complexity conception of probability. The main features of the complexity conception are:

1. It is strictly finitary: only finite sequences and finite sets of constructive objects are considered.
2. It is coding-based (which is not surprising taking into account that it is based on Kolmogorov complexity).
3. Probability distributions of the conventional probability theory are replaced by finite sets. (Which is striking if you remember that Kolmogorov was one of the founders of the axiomatic theory of probability.)
4. It is based on Condition B rather than A.

The basis of the standard statistics is the notion of *statistical model*, i.e., a family of probability distributions. Kolmogorov's suggestion was to replace statistical models with what we call in this paper *complexity models*: Classes of disjoint sets whose union contains all possible outcomes (typically data sequences) of a given trial.[2]

The main interpretative assumption of Kolmogorov's complexity conception is: we expect the realized outcome x of the trial to be *random* in A (the element of the complexity model x belong to), in the sense that

$$K(x \mid A) \approx \log |A|.$$

Formally, we can call the difference $\log |A| - K(x \mid A)$ the *deficiency of randomness* of x in A; the deficiency of randomness of x with respect to a complexity model is defined to be the deficiency of randomness of x in the element A of that model which contains x (remember that by the definition of complexity model for any x there is exactly one such A). One's acceptance of a

KOLMOGOROV'S COMPLEXITY CONCEPTION

complexity model is therefore interpreted as one's belief that the randomness deficiency of the actual outcome with respect to that model is going to be small.

At a seminar at the Moscow University (1982) Kolmogorov has given several examples of complexity models which we reproduce here; all sequences are finite.

EXAMPLE 1. A binary sequence x is Bernoulli if

$$\left.\begin{array}{l} N = \text{length of } x \\ k_0 = \text{\# of 0s in } x \\ k_1 = \text{\# of 1s in } x \end{array}\right\} x \text{ is random given } N, k_0, k_1$$

(as defined in the previous section). Formally, the Bernoulli complexity model consists of all equivalence classes, where the equivalence of two sequences means that they have the same length and the same number of 1s (and, therefore, the same number of 0s).

EXAMPLE 2. A binary sequences x is *Markov* if

$$\left.\begin{array}{l} N = \text{length of } x \\ s = \text{1st element of } x \\ k_{00} = \text{\# of 00 in } x \\ k_{01} = \text{\# of 01 in } x \\ k_{10} = \text{\# of 10 in } x \\ k_{11} = \text{\# of 11 in } x \end{array}\right\} x \text{ is random given } N, s, k_{00}, k_{01}, k_{10}, k_{11}.$$

Analogously to the previous example, the Markov complexity model consists of all equivalence classes, where the equivalence of two sequences means that they have the same length, the same first bit and the same number of transitions $i \to j$ for all $i, j \in \{0, 1\}$.

EXAMPLE 3. Markov sequences x of order d: defined analogously to the previous example.

EXAMPLE 4. A sequence x of real numbers is *Gaussian* if

$$\left.\begin{array}{l} N = \text{length of } x \\ m = \text{sample mean } \frac{1}{N} \sum_{n=1}^{N} x_n \\ \sigma^2 = \text{sample variance } \frac{1}{N} \sum_{n=1}^{N} (x_n - m)^2 \end{array}\right\} x \text{ is random given } N, m, \sigma^2$$

(to make this requirement precise in the framework of Kolmogorov's approach, one should "discretize", in some way, the real line).

EXAMPLE 5. In a similar way one can define *Poisson* sequences.

Notice that only Example 1 is of interest from the point of view of the frequentist interpretation. The other examples are natural developments, but it is clear that they go beyond frequentism.

Some mathematical results about complexity models are proven in Asarin's papers (1987; 1988). A typical result is of the following form: if x is Gaussian (see Example 4), N large and $[a, b]$ a fixed interval,

$$\frac{\#\{x_n \in [a,b]\}}{N} \approx \frac{1}{\sqrt{2\pi}\sigma} \int_a^b e^{-\frac{(t-m)^2}{2\sigma^2}} dt.$$

Notice that Kolmogorov's complexity conception actually eliminates probability from the foundations of statistics. Of course, one can reintroduce probability by identifying it with some relative frequencies; say, in the situation of Example 2 we can define the conditional probability that a 0 will be followed by a 1 in the Markov sequence x to be the ratio of the number of occurrences of 01 in x to the total number of occurrences of 00 and 01 in x. It is not clear, however, whether Kolmogorov was willing to make this step.

In general, a typical scenario of interaction between Statistician and User might look as follows:

1. User is interested in a sequence x, which might be partially (or even fully) known.
2. User gives Statistician the known part of x and other relevant information.
3. Statistician and User define a set $A \ni x$ (this set may depend on x) in which x is, they believe, random.
4. Statistician extracts a lot of useful information about the unknown part of x from the hypothesis that x is random in A. (Or, if x is known, A serves as a summary of useful information in x.)

If we forget about item 4, it becomes too easy to satisfy item 3: just take $A = \{x\}$. Item 4, when combined with Occam's razor, requires that A should be simple. This has given rise to several mathematical definitions which have been actively studied. At the same 1982 seminar at the Moscow University Kolmogorov defined a finite object x to be (α, β)-*stochastic* (where α and β are, typically small, positive integers) if there exists a finite set A such that:

$$x \in A, \quad K(A) \leq \alpha, \quad K(x \mid A) \geq \log |A| - \beta.$$

The first to study Kolmogorov's notion of stochasticity were Shen' (1983) and V'yugin (1985); they answered the question of how many (in different senses) stochastic sequences are there. There are many other interesting mathematical questions, both answered and open, raised by Kolmogorov's complexity conception; see, e.g., V'yugin (1987) and Li and Vitányi (1997), Subsection 2.2.2.

5. ALGORITHMIC THEORY OF RANDOMNESS

An in-depth review of the modern algorithmic theory of randomness is given in Li and Vitányi (1997). As already mentioned, that theory is a mixture of Kolmogorov's complexity conception of probability and Kolmogorov's axiomatics of probability. Some of the main steps of the algorithmic theory of randomness were:

- Martin-Löf (1966) defined randomness with respect to computable probability distributions. He also demonstrated the "universality" of Kolmogorov's definition of randomness restating it in terms of universal p-values.
- In 1971 Schnorr (1971b; 1971a) defined randomness through martingales (first introduced by Ville (1939) to fix the problem with von Mises's definition of collectives).
- Levin (1973) defined randomness with respect to "computable" (technically, constructively closed) classes of probability distributions. This was an important step since computable probability distributions are far too narrow for

statistics: "almost all" probability distributions in interesting statistical models are not computable. Levin also defined "universal i-values" (a Bayesian analogue of Martin-Löf's "universal p-values"). See also Levin (1976) and Gács (1980)).

REMARK 4. The algorithmic theory of randomness seems to be a natural development of Kolmogorov's complexity conception, but still it is unlikely that Kolmogorov would have approved it. Probably he would have moved in a different direction, though it is difficult to say what it would have been. He was reluctant to generalize and used to say "you are too quick" to his impatient students.

Schnorr's development was the base for Shafer and Vovk's (2001) approach to probability and finance. In (Shafer and Vovk, 2001) Kolmogorov's Condition B is replaced by the assumption that Statistician cannot win much playing a certain game.

Differences between Kolmogorov's Complexity Conception and the Algorithmic Theory of Randomness

There are a lot in common between Kolmogorov's complexity conception and the algorithmic theory of randomness; the former can be reduced to the latter and the latter can be "almost reduced" to the former. To embed Kolmogorov's theory into the algorithmic theory of randomness, the set A should be mapped into the uniform probability distribution in A. Vice versa, we can "almost embed" the algorithmic theory of randomness into Kolmogorov's theory replacing a probability distribution by its "carrier", the set where most of the distribution is concentrated (for details, see, e.g., Shen' (1983)).

However, there are important differences as well. One example has already been mentioned: the notion of Bernoulli sequence. The definition given by the algorithmic theory of randomness corresponds to the usual intuition of "coin-tossing sequences". In this subsection we consider Kolmogorov's definition from the point of view of the algorithmic theory of randomness. It can be seen that Kolmogorov's definition boils down to randomness with respect to all exchangeable distributions.

KOLMOGOROV'S COMPLEXITY CONCEPTION   63

For the binary case the difference between the two definitions was characterized in (Vovk, 1986). That paper was the outcome of my discussion with Kolmogorov (who was my Ph.D. supervisor at the moment) about the "true notion of Bernoulli sequence". It was found that the elements $\{0,1\}_k^N$ (by $\{0,1\}_k^N$ we denote the set of all sequences of length N containing k ones) of Kolmogorov's complexity model (see Example 1 above) are too fine (from the point of view of the algorithmic theory of randomness). To recover the Bernoulli sequences in the sense of the algorithmic theory of randomness within Kolmogorov's approach, his definition should be modified as follows: a finite binary sequence x is *Bernoulli* if

$$\left.\begin{array}{r} N = \text{length of } x \\ k = \# \text{ of 1s in } x \\ k^* = \lfloor \sqrt{n} \arccos\left(1 - \frac{2k}{n}\right) \rfloor \end{array}\right\} x \text{ is random given } N, k^*.$$

(Notice that the size of the set of k mapped into the same k^* has the order of magnitude $\sqrt{\frac{k(n-k)}{n}}$, i.e., $\sqrt{n}$ when k is not too close to 0 or n.)

Another result stated in (Vovk, 1986) clarifies the difference between the randomness deficiency $d_{\text{Bern}}(x)$ with respect to the class of Bernoulli distributions and Kolmogorov's deficiency $d_{\text{exch}}(x)$ (the randomness deficiency with respect to the class of exchangeable distributions). The deficiency of Bernoulliness can be split as follows:

$$d_{\text{Bern}}(x) \approx d_{\text{exch}}(x) + d(k \mid k^*),$$

where $d(k \mid k^*)$ stands for the randomness deficiency of k in the class of all k' mapped into the same k^* as k (for the exact statement of this result see (Vovk, 1986)). This implies that the difference $|d_{\text{Bern}}(x) - d_{\text{exch}}(x)|$ is at most $\frac{1}{2} \log n$. The terms $d_{\text{Bern}}(x)$ and $d_{\text{exch}}(x)$ can be as large as n, but still the difference (less than $\log n$) between them can be important: see (Vovk et al., 1999).

REMARK 5. The question about the relationship between exchangeability and iid randomness is typically answered by de Finetti's theorem and its variants (such as "finite de Finetti's theorems": see, e.g., Diaconis and Freedman (1980)); the above results can be regarded as another version of de Finetti's theorem. De Finetti's theorem itself asserts that, in the case of infinite sequences, exchangeable distributions are mixtures of iid distributions, and so there is no difference between iid and exchangeability randomness deficiency. (And so von Mises did not have to worry about the difference.) In the finite case this is only "approximately true". The probable reason why Kolmogorov chose exchangeability as his starting point is that it fits more neatly his framework; besides, it is quite close to iid.

Probably similar results can be obtained for other complexity models (such as those given in Examples 1–4) as well, but no such results are known yet.

6. CONCLUSION

The goal of Kolmogorov's complexity conception was to give better foundations for the applications of probability; Kolmogorov always believed that his 1933 axioms were sufficient for the theory of probability. The fate of the complexity conception is quite different from that of Kolmogorov's axioms: Even if understood broadly as the algorithmic theory of randomness, it is being developed by much fewer researchers and its results are rarely used outside that theory itself. In this section I will discuss limitations (responsible for its relative lack of popularity) and advantages of the algorithmic theory of randomness (as I see them). In the form of a table they are represented in Table 1. Probably the most important limitation is that neither Kolmogorov complexity nor randomness deficiency are computable. Moreover, most non-asymptotic results can be stated only "to within an additive constant", which makes them less elegant (typically even additional layers of quantifiers appear). One way to overcome this difficulty is to consider computable approximations to randomness deficiency (see, e.g., http://www.clrc.rhbnc.ac.uk). Such approximations are often given in terms of familiar statistical notions (such as tolerance intervals), but the "universal" notion of randomness deficiency gives us the clear-cut goal of approximating it. Assuming that we are given an oracle computing randomness deficiency, we can study the properties of the "ideal" versions

Advantages	Disadvantages
We can deal with finite sequences	Awkward "additive constants"
Avoiding elaborate (esp. infinitary) objects such as probability distributions	Instead of measure theory, new (and perhaps less elegant) mathematics
Simple and convincing mode of application (based on the notion of randomness) to the real world	"Probability" not defined at all
Randomness for individual sequences	Again "additive constants"; non-computability
Rich source of ideas	When ideas are developed, it is better to translate them into something different

Table 1. Kolmogorov's complexity conception vs the algorithmic theory of randomness

of our practical procedures. For example, we can prove theorems about relations between the iid and exchangeability models (de Finetti' theorem; see the previous section), or between the p- and i-versions of randomness deficiency and statistical procedures based on them (which sheds new light on the relation between sampling theory and likelihood theory).

One impediment to the development of the algorithmic theory of randomness was that simpler results of that theory go very little beyond traditional probability theory and statistics. The first reaction to Kolmogorov's complexity conception of probability was the excitement about the possibility of proving "point-wise" results (i.e., results about individual random sequences). However, it soon became clear that it is often too easy to prove such results: for example, the usual proofs of the strong law of large numbers can be easily modified to prove that the statement of the strong law of large numbers holds for all Martin-Löf random infinite sequences. Later it became clear that there is scope for non-trivial results as well, even in the case of infinite sequences: see, e.g., V'yugin (1998) or Vovk (1987).

As we already mentioned, the question of what probability is remains unanswered in Kolmogorov's complexity conception. We can follow von Mises and Reichenbach and say that, if a long binary sequence is Bernoulli, the probability of 1 is the relative frequency of 1s in that sequence; analogously, when given a long binary Markov sequence of order 2, we can identify the probability that 01 will be followed by 0 with the ratio of the number of subsequences 010 to the total number of subsequences 010 and 011. It seems that the question of what probability really is became less important eventually: Kolmogorov's complexity conception tells us how to use statistical models (in the form of complexity models) without answering this question.

Acknowledgments

This paper is based on a talk given at Roskilde University, Denmark, in the framework of a meeting on the history and philosophy of probability theory organized by the Danish Network for the History and Philosophy of Mathematics. I am very grateful to our host Prof. Stig Andur Pedersen and to the organizers of the meeting for their excellent work. The comments and criticism of the participants in the meeting are gratefully appreciated.

This work was partially supported by EPSRC through grants GR/ L35812 ("Support Vector and Bayesian learning algorithms"), GR/M14937 ("Predictive complexity: recursion-theoretic variants"), and GR/M16856 ("Comparison of Support Vector Machine and Minimum Message Length methods for induction and prediction").

Computer Learning Research Centre
Department of Computer Science
Royal Holloway
University of London
England

NOTES

[1] Estimate 100,000 can be improved to 25,550 (Bernoulli) and 6,700 (De Moivre): see e.g., Shafer and Vovk (2001).

[2] "Trial" can be arbitrarily complex, e.g. generating a long data sequence.

REFERENCES

Asarin, E. (1987). Some properties of Kolmogorov δ-random finite sequences, *Theory of Probability and its Applications* **32**: 507–508.

Asarin, E. (1988). On some properties of finite objects random in the algorithmic sense, *Soviet Mathematics Doklady* **36**: 109–112.

Church, A. (1940). On the concept of a random sequence, *Bulletin of American Mathematical Society* **46**(2): 130–135.

Diaconis, P. and Freedman, D. (1980). Finite exchangeable sequences, *Annals of Probability* **8**: 745–764.

Gács, P. (1980). Exact expressions for some randomness tests, *Z Math Logik Grundl Math* **26**: 385–394.

Kolmogorov, A. (1933). *Grundbegriffe der Wahrscheinlichkeitsrechnung*, Springer, Berlin.

Kolmogorov, A. (1938). Teoriya veroyatnostei i ee primeneniya, *Matematika i estestvoznanie v SSSR*, GONTI, Moscow and Leningrad, pp. 51–61.

Kolmogorov, A. (1951). Veroyatnost', *Bol'shaya Sovetskaya Ehntsiklopediya*, 2nd edn, Vol. 7, Bol'shaya Sovetskaya Ehntsiklopediya, Moscow, pp. 508–510.

Kolmogorov, A. (1956). Teoriya veroyatnostei, *Matematika, ee soderzhanie, metody i znachenie*, Vol. 2, Izdatel'stvo AN SSSR, Moscow, pp. 252–284.

Kolmogorov, A. (1959). Teoriya veroyatnostei, *Matematika v SSSR za sorok let*, Vol. 1, Fizmatgiz, Moscow, pp. 781–795.

Kolmogorov, A. (1963). On tables of random numbers, *Sankhya. Indian Journal of Statistics A* **25**(4): 369–376.

Kolmogorov, A. (1965). Three approaches to the quantitative definition of information, *Problems of Information Transmission* **1**: 1–7.

Kolmogorov, A. (1968). Logical basis for information theory and probability theory, *IEEE Transactions of Information Theory* **IT-14**: 662–664.

Kolmogorov, A. (1983a). Combinatorial foundations of information theory and the calculus of probabilities, *Russian Mathematics Surveys* **38**(4): 29–40.

Kolmogorov, A. (1983b). On logical foundations of probability theory, *in* Y. V. Prokhorov and K. Itô (eds), *Probability Theory and Mathematical Statistics*, Vol. 1021 of *Lecture Notes in Mathematics*, Springer, pp. 1–5.

Kolmogorov, A. and Gnedenko, B. V. (1948). Teoriya veroyatnostei, *Matematika v SSSR za tridtsat' let*, Gostekhizdat, Moscow and Leningrad, pp. 701–727.

Levin, L. (1973). On the notion of a random sequence, *Soviet Mathematics Doklady* **14**: 1413.

Levin, L. (1976). Uniform tests of randomness, *Soviet Mathematics Doklady* **17**: 337.

Li, M. and Vitányi, P. (1997). *An Introduction to Kolmogorov Complexity and Its Applications*, 2nd edn, Springer, New York.

Martin-Löf, P. (1966). The definition of random sequences, *Information and Control* **9**: 602–619.

Schnorr, C. (1971a). A unified approach to the definition of random sequences, *Math Systems Theory* **5**: 246–258.

Schnorr, C. (1971b). *Zufälligkeit und Wahrscheinlichkeit*, Springer, Berlin.

Shafer, G. and Vovk, V. (2001). *Probability and Finance: It's only a Game!*, Wiley, New York. To appear.

Shen', A. (1983). The concept of Kolmogorov (α, β)-stochasticity and its properties, *Soviet Mathematics Doklady* **28**: 295–299.

Solomonoff, R. J. (1964). A formal theory of inductive inference. Parts I and II, *Information and Control* **7**: 1–22 and 224–254.

Ville, J. (1939). *Etude critique de la notion de collectif*, Gauthier-Villars, Paris.

Vovk, V. (1986). On the concept of the Bernoulli property, *Russ Math Surv* **41**: 247–248.

Vovk, V. (1987). On a randomness criterion, *Soviet Mathematics Doklady* **35**: 656–660.

Vovk, V., Gammerman, A. and Saunders, C. (1999). Machine-learning applications of algorithmic randomness, *Proceedings of the 16th International Conference on Machine Learning*, pp. 444–452.

V'yugin, V. (1985). On nonstochastic objects, *Problems of Information Transmission* **21**: 3–9.

V'yugin, V. (1987). On the defect of randomness of a finite object with respect to measures with given compexity bounds, *Theory of Probability and its Applications* **32**: 508–512.

V'yugin, V. (1998). Effective convergence in probability and an ergodic theorem for individual random sequences, *Theory of Probability and Its Applications* **42**(1): 39–50.

EBERHARD KNOBLOCH

EMILE BOREL'S VIEW OF PROBABILITY THEORY

Did you ever try to determine the age of the captain on the understanding that you knew the height of the mainmast? In the eyes of Joseph Bertrand this problem was similar to the application of probabilities to statistical calculations concerned with molecules (Bertrand, 1899, 29f.). Emile Borel's attitude toward kinetic gas theory was by far more positive. Borel is one of the well-known founders of measure theory. It is less well-known that he published more than 50 papers and books on the calculus of probability between 1905 and 1950. He actually was an influential scholar in probabilistic thinking unifying mathematical and philosophical aspects of probability theory.

In 1934 he rightly prophesized: The unity of sciences will be re-established in an un-expected manner. The 19th century looked for this unity in the field of mechanics and of differential equations. The 20th century will find it by studying the laws of chance (Borel, 1930, 2318).

Three major works, most of all, stimulated Borel to deal with the subject: Henri Poincaré's and Joseph Bertrand's 1899 textbooks on probability theory, published in 1896 and 1899, respectively, and Poincaré's monograph "Science and hypothesis" which appeared in 1902 (Poincaré, 1896; Bertrand, 1899; Poincaré, 1902).

He emphatically defended his own standpoint in probability theory, especially against the theories of von Mises, Keynes and Reichenbach. Hence, I would like to talk about the following questions:

1. *Probability Theory and Borel's Philosophy of Mathematics.*
2. *Foundations of Probability Theory and Objections Against It.*
3. *Borel, von Mises, and Keynes.*
4. *Borel and Reichenbach.*
5. *Scientific Determinism versus Probabilistic Indeterminism.*

1. PROBABILITY THEORY AND BOREL'S PHILOSOPHY OF MATHEMATICS

In 1925, Borel actually began the first volume of his fundamental "Treatise on the calculus of probabilities and its applications" (Borel, 1925–1939) in a purely abstract manner. He assigned a number p between 0 and 1 to each event that can only be favourable or not favourable.

This first fascicle was entitled "Principles and Classical Formulas of Probability Theory". Therein he explained its mathematical theory, not being really interested in principles or logical considerations. In 1914, he said, "One may deduce absolutely rigorous, logical consequences from this definition. But whenever we want to apply these consequences to any real problem, we must substitute the concrete probability of a real phenomenon for the abstract probability. The incertitude which adheres to every concrete measure comes back again" (Borel, 1914, 92). Later on he admitted that the axiomatization of the mathematical sciences had cleared up some philosophical difficulties and had made possible some progress in the mathematical sciences. He mentioned Kolmogorov's axiomatic theory and conceded the possibility of developing the calculus of probability as a purely mathematical science without any relation to reality, comparable to n-dimensional geometry.

But all practical difficulties return if this theoretical science is applied to an arbitrary real phenomenon: The criterion is its applications. They are the true realities Borel was mainly interested in: Insurance premiums, samples obtained by biologists, phenomena observed and predicted by physicists. There is no single statement about reality whose validity can be asserted with more than probability. Hence, he took the opposite view of David Hilbert who, in 1900, had formulated the sixth problem of his celebrated list of problems: "We have to treat by means of axioms those physical sciences in which mathematics already play today an important part. These are above all probability theory and mechanics" (Barone and Novikoff, 1977/78, 125).

The key words of Borel's philosophy of mathematics were: realities, practical value, applications. For him probability theory was an applied science; essentially a social science (Borel, 1925–1939, vol. IV, 3, 60). The task of this theory was to answer to practical questions because the mathematical answers to very many practical questions are coefficients of probability. His philosophy of practical value was a philosophy of measure and evaluation, especially if we have

1. to interpret experimental results,
2. to make a decision,
3. to orientate the choices of social and economic policy.

Hence, he objected to a normatively fixed selection of questions to which we are allowed to apply probability theory, though he, too, distinguished between legitimate and illegitimate applications. Logical arguments do not prevent the statement of the problem. To refuse the answer means to fall back on a purely abstract idea, to misunderstand the character of mathematical applications. This is exactly what von Mises or Reichenbach did in speaking of meaningless problems or needless concepts of probability (Reichenbach, 1949, VIII, 376f.).

I would like to illustrate this attitude by four examples:

EXAMPLE 1. *The Method of Majorities.* The method of majorities is closely connected with the probabilistic analysis of testimony. In spite of Bertrand's and John Stuart Mill's criticism of such an application, Borel did not reject any application of probability theory to juridical problems. In such a case the possible dependence of the judges has to be taken into consideration. The method of majorities consists in considering an opinion to be practically valid, expressed by the majority, when a more or less great number of men has made known their opinion. Though numbers must constitute the basis, we must not trust too much in them because they provide no certitude. Thus the probability can be calculated when, for instance, five men choose a certain solution supposed they are elected out of 500 people who had discussed two possible solutions of the concerned problem. It is not the task of mathematics to deduce conclusions from the numbers. The society must decide whether it prefers to condemn an innocent or to acquit a guilty person (Borel, 1909a) and (Borel, 1925–1939, vol. IV, 3, 67f.)).

EXAMPLE 2. *Linguistic Problems.* Borel discussed semantic questions of the following type: How many grains are necessary in order to speak of a heap of corn? Where is the limit between a cabin and a house, between a house and a palace? If we want to define the precise meaning of a word, we have to use a great many testimonies. The relation between an affirmative and a negative statement results in an answer being a coefficient of probability. Different people may give different answers. Therefore we need, Borel said, a precise convention in order to obtain an unequivocal word definition. In questions of language as in many other areas there are only statistical truths (Borel, 1907). The influence of Poincaré's conventionalism is obvious. In his textbook of Probability theory (Borel, 1924b, 80) Borel explicitly said that a mathematical definition is a convention, but not an arbitrary convention. It is suggested a priori by the practical study of different questions and verified a posteriori by the conformity of its consequences with the observations. For a mathematician "convention" just means: the definition cannot be rigorously demonstrated. Let us consider two examples of such definitions.

(i) In 1924, Borel defined: "The probability is a ratio of the number of *favourable* cases to the number of *possible* cases if all cases are regarded as equally probable" (Borel, 1924b, 19). That is, he gave the classical definition of probability.

Borel denied that this definition implies a vicious circle on condition that we dispose of the common notion of the sense of the words "equally probable" whenever we want to define the precise mathematical sense of the word probability. The usual language must be considered as a global acquisition of every individual. This acquisition presupposes a great number of vicious circles. As we already know, Borel did not like logicians. Hence, he added: "the logicians which pretend to construct entirely logic systems without vicious circles forget that it is impossible not to use the usual language".

(ii) Discussing geometric probabilities he remarked that he considers segments only with regard to their arithmetical measure, saying: The probability that a point M lies on a certain segment PQ being a subset of the straight line AB is proportional to the length of this segment.

EXAMPLE 3. *Social Morality.* In 1908 Borel studied the connection between the calculus of probabilities and the individualistic mentality (Borel, 1908). According to him, probability is the basis of social mathematics because it serves as a bulwark against the excesses of individualistic egoism. He began with the gospel order "Love thy neighbour as thyself", and argued as follows: "The only reasonable interpretation of this maxim is: Consider your neighbour as equivalent to a fraction of yourself, taken between 0 and 1. Then we obtain coefficients of altruism or egoism. The basis of theoretical morality would be that these coefficients should not be 0. To say, these coefficients should not be negative, would mean to wish peace instead of war". For us, this example might seem to be strange. Yet, I cited it exactly for this reason. Borel repeated it in his book *On chance* in 1914 which was reprinted in 1948. Hence we can conclude that he did not explain it inadvertendly. In 1974, Maistrov (Maistrov, 1974, 242f.) heavily criticized the last two applications of Borel saying, "Such an arbitrary interpretation of probability could have arisen only as a result of the ambiguity and vagueness of this notion". Yet, he noticed that this type of argument is basic for a subjective definition of probability. And indeed, Borel actually was a moderate subjectivist. We shall come back to this point.

EXAMPLE 4. *Denumerable Probabilities.* One of Borel's most important achievements in probability theory was his introduction of denumerable probabilities. Borel considered events that depended on a denumerable set of trials. By this route he found the Zero-One law (Novikoff and Barone, 1977). He denied the existence of sets that are not denumerable. All the points that will be needed at any time in mathematical considerations will be defined by means of a finite number of words. They constitute the practical continuum that the mathematician use. He explicitly believed that the continuum will come to be regarded only as a means of studying denumerable sets which constitute the sole reality that we are capable of attaining (Borel, 1909a, 1056). Obviously, he preferred constructivistic definitions. While the standpoint of continuous probabilities is equivalent to measure theory founded by Borel, the method of random successive selections of decimal digits leads to the theory of denumerable probabilities. His favourite research subject in this regard were normal numbers, that is numbers where the frequency of every decimal digit has the limit one-tenth on condition that they are written in the decimal system. For example, it is infinitely probable that all irrational numbers are normal. Normal numbers can justify the existence of both theories, of the theory of denumerable probability and of measure theory, because we get different results according to the point of view we have. The theory of continuous probabilities can be based on axioms and definitions being analogous to those of measure theory. There are non-measurable sets. Likewise, there are sets for which the probability cannot be defined, that is, for sets whose power is greater than that of the continuum.

2. FOUNDATIONS OF PROBABILITY THEORY AND OBJECTIONS AGAINST IT

Von Mises had said: The phrase "probability of death" when it refers to a single person, has no meaning at all for us (von Mises, 1919). Borel stressed his conviction again and again that the notion of the probability of a single case is the foundation of the calculus of probability. Whenever we want to leave the axiomatic field in order to apply the probability to a real phenomenon, we have to introduce the notion of a single case.

1. Some of these evaluations might depend on purely rational deductions: Without ever casting an homogenous, regular icosahedron, we are certain that the probability of dicing an arbitrary face is $\frac{1}{20}$. This evaluation must regulate the bets we could risk with a partner with regard to a single experience made with such a die having 20 faces (Borel, 1925–1939, vol. IV, 3, 104).

2. The evaluation would be by far more complicated if we would replace the die by a straight circular cylinder: What will be the probability that it keeps lying on the bases or on the lateral surface? In the case of a coin the probability that it keeps lying on the lateral surface will be practically zero, in the case of a pencil practically one.
3. In other cases we might use results of certain experiences. These results will often but not always be statistics of frequency. We might use other observations and certain considerations.

However that may be, the probability of a single case is *subjectively* defined by the condition of a bet on an event we are willing to accept. These conditions must satisfy obvious postulates in the elementary case such as

$$p + q = 1$$

and more complicated ones in the other cases (Borel, 1925–1939, vol. IV, 3, 105). They are now called "coherence" conditions.

Already in 1924, Borel thus discussed the foundations of rational decision theory and rational betting behaviour (Borel, 1924a), two years before Frank Ramsey and eleven years before Bruno de Finetti elaborated similar ideas. Finetti gave his famous lectures "Foresight: its logical laws, its subjective sources" in 1935 at the Institute Henri Poincaré (Finetti, 1990), which had been founded partly at Borel's instigation in 1928.

If an event like the cast of dice can be repeated very often under the same conditions, the theory of repeated trials shows us that the limit value of frequency is equal to the probability. This result is a verification, not a definition.

It is not a matter of stating after the results that the mean probability of certain judgements had a certain value. It is a matter of knowing whether we can reasonably attribute a priori a determined probability to a future event, which is unique, which does not occur in a class of completely analogous events like in the case of the cast of dice.

Let us suppose we would like to estimate the evaluation of the unknown probability of a judgement enunciated by the person A: probably X will win. How can we decide whether this probability is greater or smaller than $\frac{1}{3}$? We offer a certain sum on condition that A chooses between two possibilities: Either he gets the money if a cast of dice shows 5 or 6 or he gets the money if X will win. The probability of the first event is $\frac{1}{3}$. If A believes that the winning chances of X are greater than $\frac{1}{3}$, he will choose the second possibility and vice versa.

A probability judgement must be translatable into a bet: The global success of a certain number of these bets is the only criterion of the value of the judgement (Borel, 1925–1939, vol. IV, 3, 90).

Now, let us consider objections against this theory of the probability of a single case.

Practical Objections.

(i) The agreement among judgements of several persons is not always a sufficient reason to believe in the exactness of these judgements.

For (*a*) they might be influenced by the same causes of error. (This is the well-known objection against the application of probability theory to juridical decisions), the scandal of mathematics as John Stuart Mill called it (Bertrand, 1899, XLIII, 318)), or: (*b*) They all might be unaware of certain circumstances (for example, in the case of horse-races). Borel rejected these objections (Borel, 1925–1939, vol. IV, 3, chap. 5). We have, Borel said, to distinguish between:

(*a'*): *subjective probabilities* related to a certain knowledge K included in the mind of an individual and

(*a"*): *objective* or relatively objective probabilities related to the same common knowledge K of many individuals $A, B, C...$

Our theory will be correct with regard to this knowledge K. This would not be the case if certain circumstances K' occur which completely falsify the result. Then the probability we use would have no reference to the 'real' probability.

(i) All individuals might not be especially ingenious, even the most ingenious among them might still be mediocre, so that their evaluations remain far from the truth.

The only possible answer is, Borel said, that we try to rely on sufficiently ingenious persons, for example, on well-experienced physicians if a certain disease has to be studied.

Apart from these practical objections, there are theoretical objections:

Theoretical Objections. Hans Reichenbach regarded a statement concerning the probability of a single case not as having a meaning of its own, but as representing an elliptic mode of speech: "In order to acquire meaning, the statement must be translated into a statement about a frequency in a sequence of repeated occurrences", he said (Reichenbach, 1949, 376).

In other words: In spite of the seeming contrary, all judgements are – strictly speaking – based on the study of frequencies.

Borel contradicted him: There are mental operations of a particular kind which are completely different from a simple observation of frequencies. Let us consider three examples.

EXAMPLE 5. A always assigns a probability p to the favourable event X which is greater than p' assigned by B. Hence A will always bet on X, B on non-X. We observe the frequency of the occurrences of the event X. We compare it with the mean value of the probabilities p and p'. It depends on the difference between this frequency and the mean values of p and p' whether A or B gains a prize of accuracy.

There is a great difference between the observed frequency which is related to events of very different nature and the statistical frequency which is usually considered.

EXAMPLE 6. This difference becomes still greater if one considers the case that the probability p attributed by A is sometimes smaller, sometimes greater than the probability p' attributed by B. In this case A bets sometimes on X, sometimes on non-X. The observed frequency is the global frequency of his successes, that is, of the occurrences of X or of non-X, depending on whether p was greater or smaller than p'. Such a frequency has certainly no reference to statistical frequencies because one brings together contrary events, X and non-X, the case that the patient dies and the case that he is saved.

Borel explained a method of experimental verification of the evaluation made by A of the probability of single cases. We compare them with those of another person B. Examples might be two physicians, two fans of tennis matches, two hikers in the mountains who estimate distances.

The conditions of the bet between A and B are fixed. There is the rule: A attributes p, B attributes p'.

If $p = p'$, then there will be no bet.

If $p > p'$ (or $q < q'$) then $\begin{cases} A \text{ bets on the possibility } X \\ \text{evaluated by him: } p. \\ \\ B \text{ bets on the probability non-}X \\ \text{evaluated by him: } p' \end{cases}$

Let x be the true probability of the event X. The expected mathematical value of A is

$$E = x - \frac{p + p'}{2}$$

In all cases, it is the player whose evaluation is less incorrect, who has a positive mathematical expectation (mean value of a random variable).

EXAMPLE 7. Let us assume that we are convinced by long experience that the evaluations of A are better than those of many other persons B, C, D etc. This confidence may be translated by a coefficient of probability, for example by 0,99. This coefficient is doubtless a coefficient of frequency. Now let us suppose that A attributes the probability 0,52 to a single case. We affirm with the probability 0,99 that this evaluation of A is correct. The utilisation of the frequency concerns the coefficient 0,99. It does not concern the coefficient 0,52 which does not correspond with any observed frequency. One could try to speak of cases where the same A stood up for a probability 0,52. But these cases generally will have nothing in common with the case in question and correspond to completely different circumstances. The mental operations which led A to the same result 0,52 will be likewise different. Hence, there is no simple observation of frequencies.

3. BOREL, VON MISES, AND KEYNES

In 1919 von Mises published his *Foundations of Probability Theory* which was based on the notion of collectives. It seemed to be, Borel admitted in (Borel, 1925–1939, vol. IV, 3, 81)), the most complete and most interesting attempt to establish a connection between purely theoretical probability and their applications.

Yet, the Achilles' heel of this axiomatic theory was Axiom 2: The irregularity, the randomness of such collectives, which should eliminate every subjectivity or even metaphysics in the notion of chance or of the probability of a single case. Such a conception was a complete failure in Borel's opinion.

Only in 1940 Alonzo Church gave a mathematically rigorous definition of randomness using the theory of computable functions. According to Borel, *the essential objection against the theory of collectives* and every similar theory is the following: The human mind is not able to imitate chance perfectly (that is, Mises' axiom of randomness or law of excluded gambling systems cannot be realized). To imitate chance means: to substitute any rational mechanism for the empirical method: This method consists in effecting an indefinite series of repeated trials: a thrown coin results in a series of face sides or legends. It is the same problem the American artist-composer John Cage had when he tried to eliminate every subjectivity from composing.

Let us imagine that a mathematician A affirms that he intends to construct an indefinite sequence

$$0100100010111...$$

having all properties of a sequence obtained by chance, that is, by successively drawing lots and giving the same probabilities to the figures 0 and 1. Let us suppose that A wrote down the first n figures and would like to write down the $(n+1)$ figure. There are two possibilities:

1. *A* takes into account – at least to a certain extent – the *n*-figures already written down, so that the probability of choosing 0 is *not* the same as the probability of choosing 1.

On condition that a second player *B* knows the method applied by *A*, *B* could bet with a mathematical expected value which is greater than 1 on one of the two possible results 0 or 1. Consequently, the sequence would not have the fundamental character of sequences resulting from chance. In other words, he who bets in favour of 0 or of 1 (on condition that the stakes are equal), can never have an expected value which is greater than 1.

2. *A* does not take into account the preceding figures in order to choose the $(n + 1)$ figure, conceding the figures 0 and 1 equal probabilities. This amounts to saying that the choice is made by a random drawing or by a mechanism which is analogous to the game "head or legend".

Hence Borel got the result: Human mind cannot imitate chance. This is the reason why we have to introduce the notion of probability of a single event whenever we want to leave the axiomatic field in order to apply probabilities to real phenomena.

It is a vicious circle to pretend to define such a probability by means of a given quantity which is more or less analogous to a collective. Von Mises proclaimed the opposite maxime: first the collective - then the probability (Borel, 1925–1939, vol. IV, 3, 8).

In 1921 John Maynard Keynes published his "Treatise on Probability". Borel's "A propos of a Treatise on Probability" appeared three years later (Borel, 1924a). At the very beginning, Borel raised the question why Keynes did not say anything in his very extended bibliography on the application of the theory of probability to physics. In Borel's eyes they were extremely important, related to true realities. Hence, he questioned whether for Mr. Keynes radioactive substances, properties of gases or of emulsions were no realities.

Obviously, what seemed to be the essential question for Keynes was the following subtle question: Let us take a lottery ticket. Have we to speak of the probability that we win, or of the probability of the judgement which we enunciate in declaring that we will win.

"Mr. Keynes thinks", Borel ironically said, "that important progress is made in observing that the notion of probability should be attached to a judgement, to a proposition rather than to a fact, or to an event". (Borel, 1924a, 49). Borel agreed that there might be an elliptical manner of speech: One implicitly understands the system of knowledge to which the probability statement is related. There are, however – Borel emphasized this aspect – cases where it is legitimate to speak of the probability of an event: These are the cases where one refers to the probability which is common to the judgement of all the best informed persons, who have all the information that is humanly possible to possess at the time of the judgements.

This probability could be modified by a new scientific discovery. But this is also the case for many numbers which are nevertheless called physical or chemical constants.

Hence, Borel praised Keynes for having especially clearly explained this subjective character of probability. Borel actually was a subjectivist. But he blamed Keynes for having denied the existence of objective probabilities as Poincaré and others called them. In 1926 he put it in the following way: The character of probability is essentially relative and even subjective but not exclusively (Borel, 1926, 1).

For Borel, the most important remark that Keynes made about this matter was the following: If we admit the subjective character of probability, it is no longer possible to make more precise or perfect what this probability has of imperfection or imprecision in its definition: Probability is defined in relation to a system of knowledge. Any modification of this system implies a modification of this probability. The probability is not made more perfect or more precise, but another probability is substituted for the first one.

Yet, there are difficulties concerning Keynes's conception. Let us suppose that person A possessing a certain body of knowledge K enunciates a judgement. The certain probability refers to this judgement. This probability might be evaluated by a person B, who knows A and K, but who has an opinion about A and K, called K'.

Hence, the numerical value of the probability of the judgement formulated by A depends for B on both K and K'.

Apart from this difficulty, there are two essential, yet problematic assertions of Keynes:

1. The probability $\varphi(x)$ is not necessarily numerical.
2. The set of all probability relationships cannot be arranged in a simple one-dimensional order between the two extremes of certainty of truth and certainty of falsity (Keynes, 1921, 42).

Keynes illustrated these statements by a figure which Borel simplified a bit:

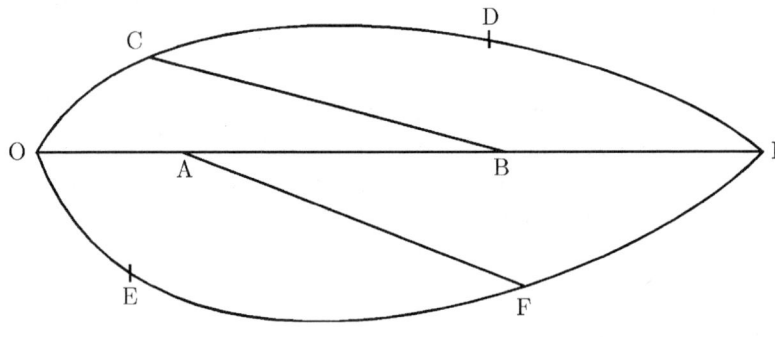

FIGURE 1

Probabilities which are numerically comparable belong to the so-called numerical path or strand, represented by $0A1$. C, D, E, F are non-numerical probabilities.

1. The two points 0 and 1 are joined by a straight line $0AB1$ and by two curves CD, EF.
2. On the straight line the numerical values are represented between 0 and 1.
3. C, D, E, F have no numerical values.
4. One can decide to say in accordance with the schema that

C is greater than 0, D is greater than C, C is less than B.

C is not comparable with A.

In the same way we could say:

E is less than F, A is less than F. E and A are not comparable.

Indeed, in the qualitative domain, it is not possible to establish a linear ordering. Take for example political parties and the tendency to range them in a linear series, from extreme right to extreme left. All that is in question is whether this complexity of judgements of quality can be truly applied to probabilities.

Basically, Borel denied this question. He compared the evaluation of prices by the method of change with the evaluation of probabilities. In spite of exceptional cases that might exist, it remains possible to establish a science of prices and exchange, of supply and demand. All things, which can be bought or sold lend themselves to a linear numerical evaluation. If one desires to know the price of coal, it suffices to offer successively greater and greater sums to the person who possesses the coal: at a certain amount he will decide to sell it. Inversely, if a possessor of the coal offers his coal, he will find it sold if he lowers his demands sufficiently (Borel, 1924a, 57).

This procedure is not falsified by certain exceptional cases.

1. There are objects which are not for sale like the Venus from Melos.
2. Sentimental reasons might hinder an owner to sell a family heirloom.
3. There are values such as the conscience of a judge which are not for sale.

In a similar way, the method of betting permits - in the majority of cases - a numerical evaluation of probabilities. Borel considered all verifiable judgements:

Somebody enunciates such a judgement, say a meteorological prediction. How can we determine or at least estimate the numerical evaluation of the probability which he attributes to his prediction?

We offer a bet on his judgement which translates so to speak, his degree of belief or conviction into a numerical estimation: We offer the choice of receiving 100 Dkk. if he is correct or 100 Dkk. if he receives a 1 or 2 in a cast of dice.

In the latter case the probability of receiving 100 crowns is one third. If the person prefers to receive 100 Dkk. if his meteorological prediction is correct, it is because he attributes to this prediction a probability superior to one third.

In a similar way as in the case of economy, there might be persons who refuse to make a bet on their health. But in spite of these exceptional cases a mathematical theory can be established which can be applied to all numerical probabilities which are evaluated. This theory will have a very large field of application.

Borel himself considered one important case not considered by Keynes where the event of which the realization is in question is not actually determined. Hence the numerical evaluation of probabilities presents some special difficulties. It can depend on judgements for which one searches to define the probabilities. Such a phenomenon is somewhat analogous to what Poincaré called nonpredicative definitions (classifications) in the theory of sets (Thiel, 1982, 148).

For example, which is the least natural number which cannot be defined by a description consisting of less than 100 Danish words? Is there such a number? It is well known that the answer is 'yes' and 'no', because the realization of the classification influences the classes coming into being.

In probability theory, we could consider the question of betting on the result of an election. Can one admit that the probability of the judgement carried by a bet has a determinate numerical value?

This is at least doubtful because the fact alone that the bet is made modifies the judgement which the bet is about:

1. The partisans of each candidate look to augment his chances by openly betting great sums on him.
2. Certain voters can be influenced by the desire to vote for the candidate who – according to their judgement – will be elected.
3. Certain betters who are at the same time voters can have their vote influenced by the desire to gain.
4. A new candidate can surge forward if the bets are large in favour of this combination.

Hence Borel drew the conclusion: If the study of certain games leads to a new chapter of the theory of probabilities, it will be a new science where psychology will be no less useful than mathematics. Von Neumann and Morgenstern continued this research. But Reichenbach opposed by saying: We must renounce a reconstruction of subjective psychological intentions (Reichenbach, 1949, 372).

4. BOREL AND REICHENBACH

Hans Reichenbach's German written "Wahrscheinlichkeitslehre" appeared in 1935, the enlarged English version in 1949. Therein he stressed four achievements of his empiricist philosophy:

1. The frequency interpretation of probability can be carried through for all uses of the term "probable". For different reasons all three interpretations of probability as
   - degree of expectation,
   - principle of indifference,
   - rational belief

   are rejected.

   Probabilities of single events as limits of frequencies have to be regarded not as *assertions*, but as *posits*. Thus the difficulties of interpreting the probabilities of single events as limits of frequencies disappear. A posit is a statement with which we deal as true, although the truth value is unknown (Reichenbach, 1949, 337). He even said, "The word 'posit' is used here in the sense as the word 'rigour' or 'bet' in games of chance". There is no need for a second concept of probability which is not reducible to frequency notions.

2. All non-deductive methods of the calculus of probability are reduced to one kind of inference: *the inference of induction* by enumeration.

3. The theory of probability has been freed from all forms of a *rational belief* in synthetic statements (philosophy of rationalism). There is no inductive self-evidence: sense observation is the only access to knowledge of the physical world.

4. The theory successfully justifies the *inference of induction* by enumeration, and therefore all forms of inductive inference.

Inductive inference is permissible because it represents an *instrument of prediction* so devised that it must lead to success if success is attainable. With regard to prediction there is no disagreement with Borel: Borel stressed the explanatory and the predicting force of probability theory as well.

Reichenbach mentioned Borel in his monumental monograph exactly once (Reichenbach, 1949, 121). For him Borel was one of the authors who introduced the concept of probability by the method of implicit definition, who used no properties of the concept other than those expressed in a set of formal relations placed as axioms at the beginning of the theory, leaving open various possibilities for its interpretation.

In other words Reichenbach numbered Borel among those authors who had introduced the concept of probability by a formal conception. Basically this is true as we saw. Yet, it is only half of the truth. He did not say any word about Borel's subjectivist attitude toward probability, he did not say any word about Borel's emphasis on the notion of probability of a single case. He did not say any word about Borel's criticism of his own, Reichenbach's theory. Perhaps too distinguished a behaviour!

Please note: No reference to Borel in spite of Reichenbach's definition of posit as bet: But he does not explain how a numerical attribution can be achieved. We posit B if $P(B) > 1/2$. But the question remains: Why do we know that? He only said: It seems to be the best we can do: "best" has a meaning that can be numerically interpreted. It refers to the posit that will be the most successful when applied repeatedly. Hence, his bet relied on the probability of repeating the bet. This is crucial.

It goes without saying that Borel especially rejected Reichenbach's theory (Borel, 1925–1939, vol. IV, 3, chap. 5). I would like to consider three arguments:

4.1. Problem of the Reference Class

If we have to find the probability holding for an individual future event, we have to incorporate the case in a suitable reference class. Reichenbach prescribed to consider the narrowest class for which reliable statistics can be compiled. Borel explained (Borel, 1925–1939, vol. IV, 3, 86) that this principle makes us consider classes in the critical case which are so far from being numerous that the frequency concept cannot be applied. In the extreme case the class will be reduced to a single case: Borel took up Reichenbach's example of a young man who has tuberculosis. The more precise the observation of the patient is, the more the difference between him and other analogous cases becomes evident. The probability that he dies is increasingly less well defined in the same measure as his case of illness is better known.

Hence, Borel frankly avowed that he admitted a part of Reichenbach's theory, but only a part. Again, he turns out not to be a dogmatist, not to be a stickler for principles.

4.2. Repetition of the Same Experience

The "frequentists" justify the evaluation "a priori" of the probability of a single case by saying that (1) the same experience can be repeated and that (2) a statistical statement can be made.

Borel rejected this justification. Only if possibility (1) is admitted, possibility (2) can occur. Hence he refuted the first possibility. His two counter-examples were:

1. the card game Bridge which requires a certain skill of the players,
2. a tennis match between two players.

A Bridge player might calculate the probability that his partners can be deceived by a feint. This probability will change when one repeats this practice playing with the same partner. It will be another probability if he should play with other partners.

The same applies to the winning chances of two tennis players who play several matches together. Both players will change tactics, so that their winning chances will not remain the same.

4.3. Reducibility to Evaluations of Frequencies

For Borel judgements of value were the most interesting type of probabilities of a single event. But this type cannot be reduced to evaluations of frequencies. It might concern material objects, individuals, groups of individuals.

Let us first consider:

1. A judgement of value which concerns a material object, say the weight of a certain suitcase. If we have to judge whether it is more or less than 15 pounds, it is a matter of a single case. We rely on earlier experiences of a rather limited number of events which does not permit to speak of a class of events. Perhaps we have lifted suitcases of which we knew the weight beforehand: 10, 12, 20 pounds. Thanks to these experiences we try to find an approximate evaluation of the weight. Our judgement has for us a certain probability.
2. A judgement of value concerning an individual: Such a judgement is necessary if we predict – for example – the result of a tennis match. We might have observed two tennis players. Even if they never played together before, we might form an opinion with regard to their chances of winning a game. This opinion can be translated into a probability coefficient related to a single event. The value of our estimation of the probability can only be verified by means of a certain number of experiences. Borel stressed the difference between such a *verification* a posteriori and a *frequency* stated a priori.

5. SCIENTIFIC DETERMINISM VERSUS PROBABILISTIC INDETERMINISM

How can we reconcile the determinism of the Laws of Nature with the indeterminism imposed by probability theory? This was Borel's leading idea when he discussed applications of probability theory to physical phenomena, like radioactive decay, kinetic gas theory, thermodynamics.

His starting point was Poincaré's statement that determinism is an indispensable postulate of scientific thinking (Borel, 1920,

2194). For the goal of science is prediction. We must rigorously suppose the observed scientific phenomena as determined.

Yet the question arises whether this necessity is absolute in the sense of mathematical truths or whether it contains a contingency, however small. Borel's answer is 'yes, it does': Physical laws are not absolutely rigorous, but rather statistical, approximate laws. *When we have to deal with reality, there is no mathematical certitude.*

Borel turns out to be an adherent of the probabilistic revolution. His solution consists in distinguishing between determinism on different scales which can be inconsistent with one another:

- between the global, rigorous, absolute determinism and
- the partial, statistical, scientific determinism.

In 1914 Borel studied two consequences of the statistical explanations of physical phenomena (Borel, 1914, chap. 6):

1. The necessity of a universal phenomenon is not incompatible with the freedom of partial phenomena;
2. The supposed absolute determinism of partial phenomena does not allow us to predict with absolute rigour the universal phenomena.

Radioactivity was Borel's example for the first consequence (Borel, 1920). The experimental laws of radioactivity prove the rigour of the global determinism of radioactive phenomena. However, there are two possible hypotheses concerning this determinism: We do not know which hypothesis has to be preferred:

1. Is it the synthesis of a great number of partial determinism? Is the development of every atom determined?
2. Or is it the statistical resultant of phenomena that cannot be individually foreseen or proven? Even the notion of causality is questionable.

Kinetic gas theory is Borel's example for the second consequence. The determinism of the molecular scale is claimed as an analogy from the human scale. But molecular determinism does not lead to determinism on the human scale (Borel, 1927, 1835).

We would have to study the movements of the molecules of the whole universe, that is an infinite number of equations that define their movements. The determinism of the global phenomenon can only be comprehended in an abstract manner. Its details cannot be foreseen. We can only foresee the most probable event. The degree of this probability must be universally identified with certainty, although this 'certainty' does not have the absolute value.

We obtain a determinism on our scale that is necessary in order to understand the world. Because of this determinism, we assume a hidden but absolute determinism of the molecular phenomena.

Borel described the probability that a noticeable variation occurs during a very short period in a space of any size from that phenomenon which is most probable with respect to statistical mechanics by a drastic example: One million monkeys pound at random one million typewriters ten hours daily for one year. The produced text contains an exact copy of all books of the richest libraries in the world. In other words: We can remove the uncertainty in applying the calculus of probability by using what Borel called the single law of chance: Phenomena with sufficiently small probabilities do not occur at any time.

Borel cited this "miracle" of the typing monkey with great pleasure. In daily life this means: We must act in all circumstances, as if events of such a small possibility were impossible (Borel, 1962, 3).

But what does it mean to say "sufficiently small"? Borel tried to define this notion resulting with different hypotheses. In 1930 (Borel, 1930) he deduced the limit 10^{-1000} saying that such probabilities must be considered as rigorously equal to zero.

Later on in 1939 Borel was more generous: He declared that any probability can be neglected on the 'cosmic' scale, that is, can be neglected universally, which is, smaller than 10^{-50} (Borel, 1925-1939, vol. IV, 3, 6-7).

Borel was especially interested in the solution of the following problem: In 1876 Joseph Loschmidt had criticised the method of explaining the irreversible phenomena of classic thermodynamics by means of reversible mechanical phenomena. Statistical mechanics might be interpreted as the study of different possibilities that can be deduced from partly undetermined data. It is a purely abstract point of view to assume a mechanical problem where the initial conditions are known with absolute precision. Even if this indeterminacy should be ever so small, the further movement of molecules becomes very undetermined in a few seconds: a huge number of different possibilities are a priori of equal probability. We do not try to determine the rigorously defined molecular, mechanical phenomena, but the most probable of all reactions. Reversibility in mechanics becomes a purely abstract fiction which vanishes at the slightest disturbance. Loschmidt's argument is not applicable because of the necessarily statistical character of mechanical explanations. The future is never completely determined - that is the principle of statistical mechanics - although we cannot speak of the indeterminism of the past.

This is the way Borel calculated the age of the captain knowing the height of the mainmast: The reader will certainly remember Bertrand's criticism of the application of probability theory to statistical calculation concerned with molecules, with which I began my explanations.

Department of History of Science and Technology
Technical University of Berlin
Germany

REFERENCES

Barone, J. and Novikoff, A. (1977/78). A history of the axiomatic formulation of probability from Borel to Kolmogorov: Part I, *Archive for History of Exact Sciences* **18**: 13–190.

Bertrand, J. (1899). *Calcul des probabilités*, Here cited from second edition dating 1907.

Borel, E. (1907). Un paradoxe économique: le sophisme du tas de blé et les vérités statistiques, *Revue du mois* **4**: 688–699. Also in (Borel, 1972, vol. IV, 2197–2208).

Borel, E. (1908). Le calcul des probabilités et la mentalité individualiste, *Revue du mois* **6**: 641–650. Also in (Borel, 1972, vol. II, 1033–1042).

Borel, E. (1909a). Les probabilités dénombrables et leurs applications arithmétiques, *Rendiconti del Circolo Matematico di Palermo* **27**: 247–271. Also in (Borel, 1972, vol. II, 1055–1079).

Borel, E. (1914). *Le hasard*, Presses Universitaires de France, Paris, the 2nd ed., published in 1948, is cited.

Borel, E. (1920). Radioactivité, probabilité, déterminisme, *Revue du mois* **21**: 33–40. Also in (Borel, 1972, vol. IV, 2189–2196).

Borel, E. (1924a). A propos d'un traité de probabilités, *Revue philosophique* **98**: 321–326. Also in (Borel, 1925–1939, vol. IV, 3, note II, 134–146), here cited from (Kyburg and Smokler, 1964, 45–60).

Borel, E. (1924b). *Eléments de la théorie des probabilités*, J. Hermann, Paris.

Borel, E. (1925–1939). *Traité du calcul des probabilités et de ses applications*, Gauthier-Villars, Paris. 4 volumes.

Borel, E. (1926). *Applications à l'arithmétique et à la théorie des fonctions*, in (Borel, 1925–1939, vol. II. 1).

Borel, E. (1927). Les lois physiques et les probabilités, *Revue scientifique* **65**: 225–228. Also in (Borel, 1972, vol. III, 1827–1837).

Borel, E. (1930). Sur les probabilités universellement négligeables, *Comptes Rendus hebdomadaires des Séances de l'Académie des Sciences* **190**: 537–540, also in (Borel, 1972, vol. II, 1139–1142).

Borel, E. (1962). *Probabilities and Life*, New York. Translated from the French by M. Baudin.

Borel, E. (1972). *Oeuvres*, Vol. 1–4, Paris.

Finetti, B. (1990). *Foresight: Its logical laws, its subjective sources*, in (Kyburg and Smokler, 1964, 93–158). English translation with supplements of the French original version: (Poincaré, 1937).

Keynes, J. (1921). *A Treatise on Probability*, Macmillan and Co, London. I cite the reprint: (Keynes, 1973).

Keynes, J. (1973). *The Collected Writings*, Vol. VIII, St. Martin's Press, New York.

Kyburg, H. and Smokler, H. (eds) (1964). *Studies in Subjective Probability*, New York, London, Sydney.

Maistrov, L. (1974). *Probability Theory, A Historical Sketch*, Academic Press, New York - London. Translated and edited by Samuel Kotz.

Novikoff, A. and Barone, J. (1977). The Borel law of normal numbers, the Borel zero-one law, and the work of van Vleck, *Historia Mathematica* **4**: 43–65.

Poincaré, H. (1896). *Calcul des probabilités, Leçons professées pendant le deuxième semestre*, A. Quiquet, Paris.

Poincaré, H. (1902). *Science et hypothèse*, Flammation, Paris.

Poincaré, H. (1937). *La prévision: ses lois logiques, ses sources subjectives*, Annals de l'Institut Henri Poincaré **7**: 1–68.

Reichenbach, H. (1949). *The Theory of Probability: An Inquiry into the Logical and Mathematical Foundations of the Calculus of Probability*, University of California Press, Berkeley – Los Angeles. English translation (of the German edition, 1934) by E. H. Hutten and M. Reichenbach.

Thiel, C. (ed.) (1982). *Erkenntnistheoretische Grundlagen der Mathematik*, Gerstenberg, Hildesheim.

von Mises, R. (1919). Grundlagen der Wahrscheinlichkeitsrechnung, *Mathematische Annalen* **5**: 52–99.

BERNA EDEN KILINÇ

THE RECEPTION OF JOHN VENN'S PHILOSOPHY OF PROBABILITY

John Venn's 1866 treatise on probability *The Logic of Chance* is taken by many commentators as the first systematic account of the frequency theory of probability. Yet, though well known and respected in Britain, Venn's work did not have a wide following in the nineteenth or twentieth centuries. There is a curious distance separating Venn from the twentieth century frequentists, such as Richard von Mises and Hans Reichenbach.[1] It appears as if the later generations did not read Venn much, or else if they did read him, they did not learn much from him. In this paper, I survey the reception of Venn's work, and examine the reason for why that reception was not wide-spread. Venn was familiar to the new generation of logicians and statisticians in Britain, such as Francis Herbert Bradley, Francis Y. Edgeworth and John Maynard Keynes, but the latter did not subscribe to the theory Venn had developed. Surveying their arguments, I show that the reference class problem was the main reason why Venn's theory did not dominate the philosophy of probability in the twentieth century.

Broadly understood, the reference class problem is the difficulty of characterizing the populations that can provide reliable statistical information. At bottom, it can be seen as the philosophical problem of characterizing the very notion of a population, and emerges in various forms and disguises in all kinds of statistical reasoning. As Stephen Stigler observed in his *History of Statistics*, this was also an issue of equally practical significance: "[t]he central conceptual problem that nineteenth-century statisticians encountered in extending statistical methodology from astronomical to social data was the isolation of social data into homogeneous classes or categories." (Stigler, 1986, 221). In order for the error analysis that was developed to analyze the deviations from a stipulated true value to apply to population phenomena, populations should have analogous true centers. But how can such true centers be understood—in reference to which boundaries?

Ironically, Venn was the first to formulate and analyze this problem with due philosophical depth. However, this was at the same time an acknowledgement of a weakness of the frequency approach, which the critics of Venn were quick to foreground. My account of these developments begins with a presentation of Venn's analysis of the reference class problem. Next I survey the reactions of a sample of Venn's readers. Examining the role of reference classes within the classical theory of probability–which was the target of Venn's philosophical position–I then argue that Venn's critics could justifiably draw upon this theory in order to deal with the problems of population thinking. I conclude by briefly contrasting Venn's approach with twentieth century formulations of frequentism, which seem to eschew the reference class problem, at least in theory.

1. VENN'S ANALYSIS OF THE REFERENCE CLASS PROBLEM

In *The Logic of Chance*, Venn expounded the view that the probability of an event was the limiting relative frequency with which that event occurred in a population of wide extent and of long duration, and positioned himself against the prevailing view of probability deriving from the French tradition, especially as perfected by Pierre-Simon Laplace–the classical theory of probability as it has been commonly called.[2] Criticizing the received tradition in probability calculus, which took the degree of certainty of a rational individual as the touchstone for measuring probabilities, Venn maintained that uncertainties could not be measured by a "belief-meter". Even had such a measuring device existed, it would presumably only indicate the vacillations of the beliefs of a single individual. (Venn, 1866, 65). Statistical frequencies, in contrast, furnished the best means of steadying individual uncertainties. Venn formulated his view in a mathematically naive manner, without invoking a rigorous notion of limit: "When we say, for instance, that it is an even chance that an unvaccinated person recovers from the smallpox, the meaning of this assertion is that in the long run each alternate person attacked by that disease does recover." (Venn, 1866, 109). His contention was that errors of estimation would be minimized when the scope of experience spread out over statistical counts.

Although the reference class problem was grounds for hesitation in the use of statistics from early on, Venn deserves the credit for the formulation of this problem with full clarity. This issue, moreover, had an unprecedented urgency for his account of probability. Venn's frequency account rested on the concept of a series consisting of homogeneous entities—coming from a natural kind–which should be presumed indefinitely extended. As Venn formulated his theory, the concept of a series involved no abstract set. Rather, it was based on the notion of a natural kind.

Venn was one of the first to employ the expression "natural kind" in English in his 1866 treatise on probability, see Hacking (1991). He did not provide a characterization of the concept of natural kind, leaving the meaning implicit in the role he accorded it in his philosophy of probability. While this usage may have been informed by John Stuart Mill's notion of kind, it has to be kept in mind that Venn had reservations about the latter. In his *A System of Logic*, Mill defined the notion of a "real kind" as a class which possessed a multitude of properties in addition to the few that were used in its characterization (Mill, 1974 [1843], Vol.7, 126). Venn thought Mill's definition was inadequate vis-a-vis historical change: "Mill, as we all know, writing in præ-Darwinian days, greatly overrated the distinctness and the ultimate or primitive character of these various attributes."[3]

In *The Logic of Chance*, Venn invoked natural kinds precisely for the purpose of delineating the populations which could exhibit statistical regularities. The uniformity of statistical enumerations, Venn noted, was "owing, much more than is often suspected, to this arrangement of things in natural kinds, each kind containing a large number of individuals" (Venn, 1866, 246).

It may be instructive here to consider an alternative world Venn imagined in his *The Principles of Empirical or Inductive Logic* where no probabilistic reasoning could be justified. If an all powerful "ingenious and malicious" agent wished to put a stop to all human inference, probabilistic or demonstrative alike, what should he do? Venn proposed the following trick:

[L]et each animal and plant and fruit, and so forth, be unique of its kind, like the fabled phænix – we might add to the number of species in proportion as we diminished the number of their representatives, so as to keep up the quantity of individuals and add to the consequent perplexity – and nearly all the generalizations and inductive extensions upon which we depend for guidance in daily life would be gone at once (Venn, 1994, 97).

If only that agent pushed Earth's orbit into a hyperbolic one, speculated Venn, so that "we should never again have any one summer or winter or day or night which would be an exact repetition of the preceding one", all probabilistic inference would be frustrated at once (Venn, 1994, 97). With the disruption of daily or yearly uniformities, never "would an average of any number afford safe guidance as to the repetition of such an average again" (Venn, 1994, 97). Natural kinds supported statistical generalizations–not necessarily for eternity, but for some stretch of time comprised within their life spans. Frequentism was not compatible with all possible worlds; it presupposed an ontology of kinds, however short-lived the latter may be.

Venn thought the existence of kinds was a sine qua non of statistics. Yet, influenced by the theories of evolution, Venn also maintained that natural kinds were not temporally enduring populations. Since no natural kind could encompass an infinity of entities existing in a finite duration of time, a series spanned an infinite time. Not only social change but also biological and geological change could not be neglected in those eons of millennia a series was supposed to extend through. As a consequence, Venn was extremely guarded in the way he related statistics to probability. The problem, as he saw it, was twofold. The first problem was the determination of the natural kinds comprehending a given individual or population. The second problem was the determination of the temporal extension of that natural kind, if any.

Venn objected basing probability judgments on statistics pertaining to a genus containing an individual, when a species could be determined which also contained that individual. The rule

Venn suggested here was to find the most homogenous population. For instance, Venn objected on the basis of this consideration the applications of probability theory to courtroom decisions. The credibility of a testimony could not be evaluated by statistical figures, Venn maintained, because each testimony was utterly unique:

> The circumstances under which the statement is made instead of being insignificant are of overwhelming importance. The appearance of the witness, the tone of his voice, the fact of his having objects to gain, together with a countless multitude of other considerations which would gradually come to light, would make any sensible man utterly discard the assigned average. He would, in fact, no more think of judging in this way than he would of appealing to the Northampton tables of mortality to determine the length of life of a soldier who was already in the midst of a battle. (Venn, 1866, 234).

Ultimately, statistical enumerations, and hence probabilistic considerations, were of no use in this case, because:

> There is not here any system of natural classification universally recognized, and appealed to as final, so that there may be general agreement as to the class by the statistics appropriate to which each party is ready to stand or fall. (Venn, 1866, 237).

A "natural" classification, one which was agreeable to all parties, and hence objective as Venn would say in his later work, was indispensable to frequentism.

A system of natural classification was a prerequisite of probabilistic reasoning, but its existence did not suffice to solve the problem of the reference class, because an individual could belong to several different classes in a hierarchical system of classification, such as the ones used in systematics. In regard to this problem, Venn distinguished between two cases, first pertaining to nested classes and the second to overlapping classes. In the first case, that is, when the several classes to which an individual belongs are successively included one within another, Venn's proposed solution was to consider the narrowest "natural kind", if there is any, that could be statistically surveyed, such that the statistics provided by that group could be presumed identical with that coming from any narrower group in the hierarchy. In Venn's example, the life expectancy of John Smith, a Suffolk farmer, can be found by reference to at least three classes: Suffolk farmers, men or animals, each of which class enlarges the previous one. In this case, both Suffolk farmers and animals could be left aside, for Venn thought:

> In such an example as the one mentioned above, where one of the classes— man— is a natural kind, there is such a complete break at this point, that on the one hand no one would ever think of introducing any reference to the higher classes with fewer attributes, such as animal or organized being. And on the other hand the inferior classes, such as farmer or inhabitant of Suffolk, do not differ sufficiently in their characteristics from the class man to make it worth our while to attend to them. (Venn, 1866, 181).

When, however, "these successive classes are not ready marked out for us by nature, and thence arranged in easily distinguishable groups", Venn could not provide a determinate rule; the selection of the group was then "more obviously arbitrary". (Venn, 1866, 182). The chance of a man's house being burnt down could be found through a survey of all kinds of buildings or only dwelling-houses or dwelling-houses with ovens, etc.; but none of these were natural kinds. Ideally, Venn contended, the "more special the statistics the better", but he conceded this was an unattainable goal. One was confronted here with an optimization problem with no optimum solution: When more attributes of an individual were taken into account, the narrowest class including that individual diminished correspondingly in size, ultimately making impossible the collection of any extensive statistical surveys. Thus, the only guiding principle Venn could offer was a practical one: "We must be guided here by the statistics which we happen to be able to obtain in sufficient numbers." (Venn, 1866, 183).

According to Venn, a more aggravated problem arose when an individual belonged to two classes which overlapped only partially. Venn's previous rule–that is, that the narrowest homogeneous natural class should be given priority in the determination of the reference class–could not be applied here, because the two classes were not nested into each other. And a devastating problem for the determination of probabilities emerged when the two classes indicated different statistical frequencies. In Venn's example, a visit to Madeira had adverse effects on the health of an average Englishman (nine out of ten Englishmen suffered from various health problems during their stay there), whereas such a visit was beneficial to anybody suffering from consumption (nine out of ten consumptive men recovered when they visited the island).[4] John Smith was a consumptive Englishman. Venn asked: "[A]re we to recommend a visit to Madeira in his case or not?" (Venn, 1866, 183). This was a case where statistics yielded two different probabilities to John Smith's benefiting from his visit to Madeira. "We should here have two sets of statistics, natural kinds they might almost be called, which would offer decidedly different results." (Venn, 1866, 230). Could there be two different objective probabilities for the happening of one and the same event? If not, which statistics was to be preferred?

Venn maintained once again that the ideal solution would be to obtain statistics from the smaller subclass, consumptive Englishmen. Yet, lacking that statistical information, or worse, not being able to survey the long term behavior of such a small population, the problem was once again deferred to good judgment and common sense. Venn's observation that longevity had more to do with the state of health of an individual rather than with his nationality could be applicable to this situation, but there was no context-independent solution to the general problem. Insurance companies, Venn conceded, could make use of either of the two statistics.

Whether the classes to which an individual belonged were encaptic or not, the problem of determining the unique class, that is, the series in Venn's terminology, posed a real threat to the objectivity of the frequency accounts of probability. The existence of a natural and universal system of classification, or the existence of natural kinds, could alleviate the difficulty, but only to some extent. No matter how enthusiastically Venn embraced frequentism, he could not conceal his perplexity on this issue. There were, as a rule, a plenty of series an individual event could be referred to. Venn observed: "It is obvious that every individual thing or event has an indefinite number of properties or attributes observable in it, and might therefore be considered as belonging to an indefinite number of different classes of things." (Venn, 1866, 175-6). The frequency view of probability could not trim down the multiplicity of averages, arising with the multiplicity of classificatory choices.

As Venn noted, the existence of natural kinds grounded the relation between an individual and the series to which it belonged. The prospects of an individual member of a series having a given attribute, such as John Smith's probability of dying at a given age, were intimately tied to the properties of the natural kind, man, with which the individual shared some essential attributes. Yet, as has been indicated, Venn's natural kinds were not atemporal entities–they could have histories as well. The longevity of mankind, for instance, was most likely different in different epochs of history. Statistics of death from smallpox or accusations of witchcraft showed how unrepresentative such figures would be for long term tendencies. Venn was emphatic that regularity over the course of a couple of decades was no guarantee of uniformity in the long run:

> We shall, in the vast majority of instances ... find that the numerical proportions, which by their persistence produce the uniformity, gradually change, and this to such an extent that the term uniformity at last becomes inappropriate. (Venn, 1866, 24)

Venn saw neither nature nor society as the source of vast uniformities. So the reference class problem arose in a new disguise:

> If the type were fixed we could not have too many statistics, but if it vary, our extra labour may be worse than wasted. The danger of stopping too soon is easily seen, but in avoiding it we must not fall into that of going on too long.[5]

Statisticians were thus faced with a fundamental dilemma. As Venn put it, "whilst we may fall into error by taking too few instances we may also fail in our aim ... by taking too many." (Venn, 1866, 16) But what was too few and what was too many? Again, Venn did not provide a theoretical answer, only leaving the decision to the good judgment of the practicing statistician:

> [T]he limits within which we collect our statistics are to a certain extent arbitrary; we must exercise our judgment in deciding where we will draw the line and what we will include within it. (Venn, 1866, 24)

On a view of the world as an open-ended process of change, this pragmatic attitude was the best a philosopher could recommend to statisticians.

On the other hand, for mathematical purposes Venn recommended to have recourse to some form of idealization. Instead of a real series developing in actual time, he proposed considering a "substituted series", imagined to extend a finite segment of the real series. An ideal series was to remedy the defects of an actual series–the latter was possibly inhomogeneous, whereas the ideal series "must be regarded as indefinitely extensive in point of number or duration." (Venn, 1866, 19). The calculations concerning games of chance, for instance, had their reference to a "substituted series". One reasoned, "not from the fragment given to us; from potential, therefore, not from actual experience." (Venn, 1866, 298). However, Venn did not confine the mathematics of probability entirely to this ideal series. He dismissed Bernoulli's theorem, on the grounds that its presuppositions did not match with the reality of natural kinds of finite longevity.[6]

2. VENN'S RECEPTION

Venn's most enthusiastic follower was the American philosopher and scientist Charles Sanders Peirce. Peirce had formulated a frequency account of probability in an 1866 lecture, before he read Venn's *The Logic of Chance*: "A probability is a matter of fact; namely, a frequency of a species of events relatively to its genus." ("Lowell lecture" in (Peirce, 1982, Vol.1, 404).) In his 1878 "The Doctrine of Chances", Peirce acknowledged Venn:

> The conception of probability here set forth is substantially that first developed by Mr. Venn, in his *The Logic of Chance*. Of course, a vague apprehension of the idea had always existed, but the problem was to make it perfectly clear, and to him belongs the credit of first doing this. (Peirce, 1957, 64).

Peirce was more emphatic than Venn that probability judgments did not pertain to the occurrence of single events, but to the numeric relation between two kinds of events:

> "Probability," in the untechnical sense, is therefore a vague word, inasmuch as it does not indicate what one, of the numerous subordinated and coordinated genera to which every argument belongs, is the one the relative frequency of the truth of which is expressed. It is usually the case, that there is a tacit understanding upon this point, based perhaps on the notion of an infima species of argument. But an infima species is a mere fiction in logic. And very often the reference is to a very wide genus.[7]

Firmly convinced that probability was a relational term the relata of which must be made explicit, Peirce found Venn's treatment of the reference class problem inadequate. Confronted with different rates of mortality among the consumptive men and Englishmen who visited Madeira, insurance companies would not be best advised to adjust their rates to one or the other class. Rather, Peirce maintained, they should seek to determine the relational term applicable to this situation, the rate of mortality among the consumptive Englishmen. By allowing arbitrary choice in this matter, thought Peirce, Venn had fallen into "some conceptualistic errors of his own", (Peirce, 1867, 319). (See also (Peirce, 1982, Vol.2, 22–3).)

Later, at least by the year 1910, Peirce changed his opinion about the adequacy of the frequency definition of probability. He was puzzled over defining the ratio between two infinite magnitudes, over the reference to an indefinite unactualized future run of trials, and over assuming fixed probabilities in the long run. He warned that "after the probability has been ascertained, we must remember that this probability cannot be relied upon at any future time unless we have adequate grounds for believing that it has not too much changed in the interval." (Peirce, 1957, 75). How could a probability which is a limiting relative frequency change in time? Peirce's suggestion in this context was to define probability as a dispositional property of the physical set-up–as a "would-be" or "habit" of objects. This shift of opinion seems to be a way of reconciling with temporality. The generic event that "if a die be thrown from a dice box it will turn up a number divisible by three" had a probability one-third because of a property of the die in question: "The statement means that the die has a certain 'would-be'; and to say that a die has a 'would-be' is to say that it has a property, quite analogous to any habit that a man might have."[8] Peirce was still concerned, however, whether there could be different long run manifestations of such habits, and he left the issue unresolved.

Venn's *The Logic of Chance*, though widely known and respected in Britain, did not have a wide following in the nineteenth or twentieth centuries.[9] Venn was familiar to the new generation of logicians and statisticians in Britain, Francis Herbert Bradley, Francis Y. Edgeworth and John Maynard Keynes, but the latter refused politely the theory Venn had developed. The logician and philosopher Bradley objected to Venn's theory in his 1883 *Principles of Logic*:

> Mr. Venn, for whose powers I feel great respect, and from whose *Logic of Chance* we all can learn, holds that in the long run every chance will be realized. ... His book is much injured by this terrible piece of bad metaphysics. He has translated a mathematical idea into a world where it becomes an absurdity. (Bradley, 1883, 2134–214).[10]

Venn's injurious metaphysics, according to Bradley, lay in his conception of a series. Was the series a real one or a fictitious one? If real, it could not be an infinite series. Bradley conceded that infinity could be an object of study within mathematics, but outside mathematics "an infinite number is an idea that attempts to solder elements which are absolutely discrepant. It could not exist until the world, as known in our experience, was utterly shattered and transmuted from the roots." (Bradley, 1883, 213). A finite series, on the other hand, did not do the job. Finite relative frequencies were compatible with different underlying probabilities: "It is not found in experiment that actual runs do always, or often, correspond exactly to the fractions of the chances. That correspondence is after all the most probable event, but to make it more is a fundamental error." (Bradley, 1883, 212–3). Bernoulli's theorem did not justify that a series, even an infinite one, would manifest the chances precisely. Bradley's reading of Bernoulli's theorem was close to its mathematical formulation: "It is false that the chances must be realized in a series. It is however true that they most probably will be, and true again that this probability is increased, the greater the length we give to our series." (Bradley, 1883, 214). But that meant probabilities could only emerge in that "fictitious" long run Bradley could not conceive of. According to Bradley, the sole grounds for invoking that fictitious or "imaginary" series was to guarantee the causes fixed, but was this justifiable? He thought not: "In an actual fresh case I do not know the fresh conditions, and, if I did, I do not know what the old cause specially is. I do not know the actual cause (or causes) of the former series. I do not know that these are present again in the unknown case." (Bradley, 1883, 211).

Bradley's conclusion was "[i]t is mere misunderstanding which supposes that chance involves a series, and that the logic of probability is essentially concerned with statistical frequency." (Bradley, 1883, 209). He defended the classical definition of probability, as a degree of rational belief, and argued that a divorce between objective and subjective probabilities was unacceptable:

> Any theory which calls the doctrine of chances merely "objective," or merely "subjective," is certainly false. It is a vicious alternative which, if it were sound, would upset general results we have found to be true, and which is contrary to the special facts of the case. (Bradley, 1883, 202).

Judgments of mathematical probability were both objective and subjective, objective because the theory prescribed how background knowledge should bind probabilistic judgments, and subjective because such judgments did not pertain immediately to facts, but to expectations about the happening of facts.

Edgeworth was more sympathetic to Venn's viewpoint.[11] He concurred with Venn that probability theory had to incorporate large-scale phenomena that could be analyzed with statistical techniques. The primary standard for estimating degrees of probability was statistical uniformity, "the fact that a genus can very frequently be subdivided into species such that the number of individuals in each species bears an approximately constant ratio to the number of individuals in the genus." (Edgeworth, 1884a, 6). Yet he did not change his conception of probability on that account, and countered Venn by defending the classical probabilists: "on the axioms of Probabilities which the inquiry has brought under our notice, the mathematical founders of the science are better authorities than the acute logician who has attempted to undermine their work." (Edgeworth, 1884b, 28).

Unlike Venn, Edgeworth defended the principle of indifference and its use in the inverse inference. He did not think the principle was necessarily an expression of ignorance. It could be conceived of as an expression of a different kind of experience than the one Venn had envisioned–still "material" experience, Edgeworth suggested echoing Venn, but "of a more diffused sort of matter". A priori, or "intellectual" probabilities as Edgeworth referred to them, did not live in a "dreamland", but in a "more general sort of experience". (Edgeworth, 1884a, 12). However fluctuating or unreliable the strength of our impressions of such general experience may be, Edgeworth found them reliable in the average: "Peter and Paul betting against each other about an event, the chance of which is really even, are each ready to give odds. Upon an average their opposite errors counterbalance each other." (Edgeworth, 1884a, 8).

The economist Keynes lauded Venn's approach only because "[t]here is no mystery about it–no new indefinables, no appeals to intuition." (Keynes, 1921, 94)[12] Keynes' criticism was that Venn's approach restricted the theory beyond recognition: "The identification of probability with statistical frequency is a very grave departure from the established use of words; for it clearly excludes a great number of judgments which are generally believed to deal with probability." (Keynes, 1921, 95). According to Keynes, even Venn needed recourse to other grounds of probable judgment than the one afforded by statistical frequencies:

> He [Venn] forgets also that, when he comes to consider the practical use of statistical frequencies, he has to admit that an event may possess more than one frequency, and that we must decide which of these to prefer on extraneous grounds. The device, he says, must be to a great extent arbitrary, and there are no logical grounds of decision; but would he deny that it is often reasonable to found our probability on one statistical frequency rather than on another? (Keynes, 1921, 98).

The reference class problem could not be simply put under the rug. "We cannot be content with the only counsel Venn can offer, that we should choose a frequency which is derived from a series neither too large nor too small." (Keynes, 1921, 98–99). Unlike Venn's frequency theory, Keynes believed his logical approach to probability could incorporate a variety of considerations informing probabilistic judgments, such as relevance, dependence or independence.[13]

Despite his pointed criticism, Keynes revered Venn. In 1921 he sent a copy of his newly appearing treatise on probability to Venn, along with a letter signed "in a spirit of piety to the Father of this subject in Cambridge." Keynes wrote:

> It is now no less than 55 years since the appearance of your first edition; yet mine is the systematic Treatise on the Logic of the subject, next after yours, to be published from Cambridge; nor, so far as I know, has there been any such treatise in the meantime in the English language. Yours was nearly the first book on the subject that I read; and its stimulus to my mind was of course very great. So, whilst you are probably much too wise to read any more logic (as I hope I shall so in my old age), I beg your acceptance of this volume, the latest link in the very continuous chain (in spite of difference in opinion) of Cambridge thought.[14]

Keynes' letter indicates the dearth of systematic treatments of the topic after Venn. There is no indication that an enthusiastic group of followers took up and developed Venn's philosophy of probability.

Venn was too old to be involved in controversy at the time Keynes wrote him. He had actually given up his studies in logic and probability a long time ago. Since the 1880s Venn's interests became more historical than logical. He started to work on a history of Gonville and Caius College by collecting the biographies of the members of the college. The result was the massive *The Biographical History of Gonville and Caius College* (1897-1901). After he became President of Gonville and Caius College in 1903, Venn also published a family history, *Annals of a Clerical Family* (1904), tracing the lives of the nine generations of Venns, all clergymen of the Church of England.

3. REFERENCE CLASSES OF THE CLASSICAL PROBABILISTS

The reference class problem appears to be the main reason why Venn's work was not whole heartedly accepted by his followers. Confronted with the problem of historical change which dissolved timeless average men, Venn's only solution was to consider ideal series. Classical probabilists, however, were better equipped to deal with historicity. Siméon-Denis Poisson had already conceived of chance as a variable in time when he formulated his version of the law of large numbers in the 1820s. (See Poisson (1835, 1837).) Accordingly, the assumption of constant causes was neither a prerequisite nor a consequence of statistical regularity. Adolphe Quetelet introduced a tri-partite distinction among causes in his 1846 *Letters on the Theory of Probabilities*: constant, variable and accidental. What was new in his list when compared with the earlier accounts of probabilists was the variable cause. Quetelet took them to be the ones which "act in a continuous manner, with energies and tendencies which change either according to determined laws or without any apparent law." (Quetelet, 1849, 107). Constant causes had persistent and uniform effects, like gravitational attractions between the sun and the planets. (Quetelet, 1848, 288). If all causes were of this kind, however, the world would be deadly still; no life or real change would have come into existence. (Quetelet, 1849, 142–3). Quetelet intimated that the variable and accidental causes moved the universe. Variable causes, in Quetelet's terminology, could have periodic effects, like the ones encountered in meteorological phenomena, or progressive effects, like the state of art in sciences, but they could also have highly irregular effects, like "the years of abundance or scarcity" in economic phenomena. (Quetelet, 1848, 289). If the variable causes were periodic, then large-scale compilations of statistics would provide a reliable representation of their average effect, as they would erase the cumulative effect of accidental causes altogether. But if the variable causes were not periodic, then there was no guarantee that statistical summaries would indicate the dominating effect of constant causes and only those.

Quetelet's introduction of variable causes posed a problem for interpreting statistical averages, but this was not a conceptual problem of the sort that confronted frequentism. Quetelet reckoned probabilities as pertaining to single occurrences of events. On the other hand, Quetelet did not develop a quantified analysis of causes. The statistical methods known as time-series analysis, which proceed upon a decomposition of a temporal sequence of observations into the trend, cyclical variation and irregular variation, were developed in the beginning of the twentieth century.[15]

Quetelet's division among causes, suggestive of the three-fold decomposition of time-series analysis, testified to the complexity of the problem of interpreting statistical compilations. Quetelet had important insights into understanding that complexity. Statistical activity, in Quetelet's opinion, could provide only a frozen picture of society in its historical movement:

> If we consider a state during one of its phases of development, in taking its statistics, we in a manner arrest its march in order to study it more at ease, and to discover its organization and its relations with all which surrounds it. (Quetelet, 1849, 176).

Quetelet drew a division of labor between statistics and political history. Statistics to political history was, in his opinion, like statics to dynamics. "Generally, statistics relate to the present, leaving the past to history, and the future to politics." (Quetelet, 1849, 176). He did recommend, however, that statistical facts should be collected over some suitable period in order to obtain a more faithful representation of the temporary static state of the society:

> We should be wrong, however, in supposing this halt and this examination made by the statist as reduced to an infinitely short instant of time. The examination should, on the contrary, extend over a period long enough to eliminate accidental causes: We must, however, take care that it does not extend over so lengthened a period of time that the social state may have been sensibly changed during the interval. (Quetelet, 1849, 176).

Quetelet's caution about the significance of statistics within a changing society bore a remarkable similarity to Venn's. However, Quetelet did not need to be alarmed as much as Venn, since he did not advocate a frequency account of probability. For the classical probabilist, who did not identify probability with statistical frequencies, the problem posed by the variability of causes was not a fundamental threat, at least as far as the conceptual foundations of the calculus was concerned.

Like the political economists from Robert Malthus to Stanley Jevons, Edgeworth and Keynes, Quetelet had a dynamical conception of populations. In Malthus' epoch-making theory of population, demographic phenomena were anything but static. Malthus conjectured an oscillatory movement with respect to "happiness", due to the non-linear relationship between population growth and economic processes. In the second edition of *An Essay on the Principle of Population* in 1803 Malthus noted, concerning the state of happiness:

> The times of their vibration must necessarily be rendered irregular from the operation of many interrupting causes; such as, the introduction or failure of certain manufactures; a greater or less prevalent spirit of agricultural enterprise; years of plenty, or years of scarcity; wars, sickly seasons, poor laws, emigration, and other causes of a similar nature. (Malthus, 1803, 94).

The irregularity of the vibrations of happiness, trade cycles and economic crises was a constant concern for political economists in the nineteenth century. Jevons proposed in 1862 that each and every variety of periodic fluctuation in commerce should be sought out so that "we can correctly exhibit those which are irregular or non-periodic, and probably of more interest and importance."[16] Jevons himself was fascinated by a possible correlation between trade cycles and "solar variations" (the increase and decrease of sun-spots), and devoted considerable time and energy at confirming it.[17] He observed in his *Principles of Science* the intricacy of analyzing periodicity:

> In studying, then, a phenomenon of rhythmical character we have a succession of questions to ask. Is the periodic variation uniform? If not, is the change uniform? If not, is the change itself periodic? Is that new period uniform, or subject to any other change, or not? and so on ad infinitum. (Jevons, 1874, Vol.2, 64).

Jevons was not sympathetic to Venn's philosophy of probability and subscribed to the classical theory. The frequency view of probability, glossing over periodic and non-periodic change, idealizing away temporality, was not the most congenial measure of uncertainty, especially for the theoreticians of social sciences.

4. CONCLUSION

This reception story indicates a gap between the rise of statistical practice and the frequency theory of probabilities. As recent works on the history of statistical reasoning, Theodore Porter's *The Rise of Statistical Thinking* and Ian Hacking's *The Taming of Chance* document, there was a rapid and massive spread of statistical practice in the nineteenth century. A burst of statistical activity in this period, what Ian Hacking (1990, vii) has called an "avalanche of numbers", indicated an ever widening range of mass phenomena which seemed to occur in regular intervals. The notion of a statistical law became conceivable by the 1840s when those statistical regularities were projected beyond the confines of space and time (as well as paper and effort) encompassed by statistical tables. Demographers and their readers believed that they were confronted not simply with statistical facts, as for instance the kind that would be found through one single census, but with statistical laws, *i.e.*, long term regularities of aggregate phenomena. By the mid-nineteenth century, statistical practice transformed moral sciences by presenting a new picture of society as the static or dynamic state of a massive entity that was the sum total of its members. (See Daston (1987), Porter (1986), Hacking (1990) and Hilts (1973)). Yet, as Venn's concerns indicate, the frequency account of probability was not simply a theorizing of the concept of statistical uniformity brought about by this boom of statistical activity. The suggestion to the contrary, advanced by Hacking and Porter, needs to be qualified.[18]

While Hacking and Porter describe a story of expansion, my history of frequentism is not a triumph story. At the turn of the century, frequentism was only one of the several acceptable accounts of probability. My account does not conflict essentially with that of Porter or Hacking, but accentuates the difficulties of converting statistical practice into a frequency theory of probability. One of the major problems injuring that conversion concerns the implicit ontology of the frequency theories. Accordingly, probabilistic reasoning rests on the existence or postulation of large populations of similar entities. This is an implicit reference to universals, expressed in the nineteenth century through the wide spread locution of terms such as "species", "genus", "kind" or "natural class".

This aspect of frequentism was obscured in the twentieth century by the use of set theory. The frequency theory of Richard von Mises, for instance, was based upon set theoretic considerations for characterizing a random sequence. In contrast, Venn's starting point was a search for a natural substratum which can ground statistical assertions. The identification of a natural kind as the stable core of a population concerned, in Venn's approach, not only the applications of the theory, but also its conceptual framework. Twentieth century accounts of frequentism, in contrast, were written in an abstract space without much concern for this ontology of kinds. Hence the eclipse of Venn.

BERNA EDEN KILINÇ

Department of Philosophy
Boğaziçi University
Turkey

NOTES

[1] Wesley Salmon believes Hans Reichenbach did not have any first-hand knowledge of Venn's work (Salmon, 1980).

[2] John Venn (1834-1923) entered Gonville and Caius College, Cambridge in 1853, and received a degree in mathematics in 1857. His interest in mathematics waned as soon as he graduated from Cambridge; in 1857, in a fit of reaction against his narrow mathematics education in Cambridge, he sold most of his mathematics books, and began his career as a clergyman. He was ordained curate in 1858 and priest in 1859. His interest in probability theory was awakened by his reading works in history and political philosophy, in particular, those of Henry Thomas Buckle and John Stuart Mill. Conceived in 1858, Venn's work in probability theory, *The Logic of Chance*, appeared in 1866, and in later editions in 1876 and 1888. Venn resigned from his position as a priest in 1862 and was offered a lecturer position at Caius College. In this position, Venn lectured in ethics, political economy and logic. Only towards the end of 1860s could he specialize his teaching exclusively to logic. For biographical information on Venn, including an account of his unpublished autobiography, see my dissertation, (Kılınç, 1997), and (Kılınç, 1999).

[3] (Venn, 1994, 82–83). Mill's reluctance to incorporate the Darwinian point of view was also noted by Leslie Stephen, who found it curious that Mill did not change his notion of kinds in the last edition of *A System of Logic* in 1872, "after the first Darwinian controversies" (Stephen, 1950, Vol.3, 130). On Mill's notion of kinds, see Hacking (1991), Kılınç (1997) and Kılınç (2000).

[4] These are Venn's figures.

[5] (Venn, 1866, 38). See (Kılınç, 1999) for the impact of the theory of evolution on Venn's concern for historicity.

[6] See (Kılınç, 1997) and (Kılınç, 1999) for a detailed analysis of how Venn objected to Bernoulli's theorem, a version of which we call nowadays the weak law of large numbers.

[7] Peirce, "On an Improvement in Boole's Calculus of Logic" from 1867 in (Peirce, 1982, Vol.2, 22). In Peirce's idiosyncratic usage, an argument to infer that "x is an A" belonged to the genus B, if a premise of the argument was "x is a B".

[8] Peirce's supplement, dating from 1910, to his 1878 article.

[9] An exception is the mathematician George Chrystal, who had a high regard for Venn's works. See his (Chrystal, 1892) and (Chrystal, 1886-1889, Vol.2, Ch.36).

[10] Bradley (1846-1924), a graduate of Oxford, was professor at Merton College, Oxford, for the rest of his life. Known as one of the most prominent representatives of idealist Hegelian philosophy in Britain, Bradley was famed with his criticism of John Stuart Mill's philosophy of experience and ethical individualism. His *Ethical Studies* (1876) and *Appearance and Reality* (1893) constituted a target for the philosophies of Bertrand Russell and G. E. Moore.

[11] Venn suggested Edgeworth's name as a candidate to the Royal Society. As he wrote to Galton, "I have long had a very high opinion of his work." (University College London, Galton Papers, letter on November 27, 1889.) After the subject "Theory of Statistics" was introduced into the Schedule of the Moral Sciences Tripos in 1890, Venn proposed once again Edgeworth: "If it were desirable to go into the mathematical foundation of the various rules no one would do it better than Edgeworth." (University College London, Galton Papers, letter to Galton on February 23, 1890.)

[12] Keynes (1883-1946) graduated from King's College, Cambridge in 1905 writing a fellowship dissertation on the theory of probability. Afterwards, he turned to the study of economics with Alfred Marshall, and in 1908 obtained a lectureship in economics in Cambridge. His career in economics and policy making was an outstanding one, marked by epoch-making works such as *A Treatise on Money* (1930) and *General Theory of Employment, Interest and Money* (1936).

[13] Venn was criticized on similar grounds by the logician W.E. Johnson, who was a pupil of Venn and a professor of Keynes. His approach to the foundations of probability was published post-humously Johnson (1932a,b,c).

[14] (Keynes' letter to Venn from 31 August 1921). Gonville and Caius College, Cambridge, Venn Papers, C49.

[15] Time series analysis originated in the works of Arthur Schuster and G.U. Yule, in their analysis of the variations in the cycles and amplitudes of the sunspots. See (Davis, 1941, Ch.1).

[16] Jevons, "Periodic Commercial Fluctuations", 4. This paper was presented to the Meeting of the British Association at Cambridge in 1862.

[17] Taking solar variation to have an approximately eleven-year period, Jevons analyzed the variation of the prices of several agricultural products over eleven-year cycles, and thought he could thereby recover a striking pattern in that variation. He relied for this purpose on James E. Thorold Rogers' *History of Agriculture and Prices in England*, which contained a compilation of the prices of various grains in England from 1259 to 1793. See his "Solar Period and the Price of Corn", a paper he read at the meeting of the British Association at Bristol in 1875. Jevons' was not a successful attempt, for many political economists doubted of the existence of regular periodic commercial cycles in the first place. See (Morgan, 1990, Ch.1).

[18] Hacking (1990, 208) and Porter (1986, Ch. 3), suggest that the frequency account of probability was prompted by the expanding statistical activity in the period.

ARCHIVAL SOURCES

- Special Collections at the University Library, The University of Birmingham. Church Missionary Society Archives. Venn Papers.
- The Library of Gonville and Caius College, Cambridge University. The John Venn Papers.
- University College London. Galton Papers.

REFERENCES

Bradley, F. (1883). *The Principles of Logic*, Kegan Paul, Trench, &Co., London.

Chrystal, G. (1886-1889). *Algebra*, Vol. 1, 2, Chelsea, New York.

Chrystal, G. (1892). On some Fundamental Principles in the Theory of Probability, *Transactions of the Actuarial Society of Edinburgh (New Series)* **2**: 421–439.

Daston, L. (1987). Individuals versus Laws of Society: From Probability to Statistics, in *The Probabilistic Revolution, Vol. 1: Ideas in History*, pp. 295-304.

Davis, H. (1941). *The Analysis of Economic Time Series*, Indiana.

Edgeworth, F. (1884a). The Philosophy of Chance, *Mind* **9**(34): 223–35. Reprinted in (Edgeworth, 1996, Vol. 1, 6–18).

Edgeworth, F. (1884b). Chance and law. Reprinted in (Edgeworth, 1996, Vol. 1, 19–28).

Edgeworth, F. (1996). *Writings in Probability, Statistics and Economics*, Vol. 1, Edward Elgar Publishing.

Giere, R. and Westfall, R. (1973). *Foundations of Scientific Method: The Nineteenth Century*, Bloomington, Indiana.

Hacking, I. (1990). *The Taming of Chance*, Cambridge, England.

Hacking, I. (1991). A tradition of natural kinds, *Philosophical Studies* **61**: 109–126.

Hilts, V. L. (1973). *Statistics and Social Science*, in Giere and Westfall (1973).

Hintikka, J., Gruender, D. and Agazzi, E. (1980). *Pisa Conference Proceedings*, Vol. II.

Jevons, W. (1874). *The Principles of Science: A Treatise on Logic and Scientific Method*, Vol. 1, 2, Macmillan and Co., London.

Johnson, E. (1932c). Probability: The Deductive and Inductive Problems, *Mind* **41**(164): 409–423.

Johnson, W. (1932a). Probability: The Relational of Proposal to Supposal, *Mind* **41**(161): 1–16.

Johnson, W. (1932b). Probability: Axioms, *Mind* **41**(163): 281–296.

Keynes, J. (1921). *A Treatise on Probability*, Harper, New York.

Kılınç, B. (1999). John Venn's Evolutionary Logic of Chance, *Studies in History and Philosophy of Science* **30**(4): 559–.585.

Kılınç, B. (2000). *Robert Leslie Ellis and John Stuart Mill on the One and the Many of Frequentism*, in British Journal for the History of Philosophy, **8**(2): 251–274.

Kılınç, B. E. (1997). *The One and the Many of Frequentism*, PhD thesis, University of Chicago.

Malthus, R. (1803). *An Essay on the Principle of Population*, 2nd edn.

Mill, J. (1974 [1843]). *A System of Logic, Ratiocinative and Inductive: Being a Connected View of the Principles of Evidence and the Methods of Scientific Investigation*, Vol. 7, 8 of *Collected Works of John Stuart Mill*, University of Toronto Press, Toronto. Edited by J.M. Robson.

Morgan, M. (1990). *The History of Econometric Ideas*, Cambridge University Press, Cambridge.

Peirce, C. (1867). Venn's Logic of Chance, *North American Review* **105**: 317–321.

Peirce, C. (1957). *The Doctrine of Chances*, in (Wiener, 1957). Originally from 1878.

Peirce, C. (1982). *Writings of Charles Sanders Peirce: A Chronological Edition*, Indiana University Press, Indiana. Edited by Christian J.W. Kloesel.

Poisson, S.-D. (1835). Recherches sur la probabilité des jugements, *Comptes rendus hebdomadaires des séances de l'Académie des Sciences* **1**: 473–494.

Poisson, S.-D. (1837). *Recherches sur la probabilité des jugements en matiére criminelle et en matiére civile*, Paris.

Porter, T. (1986). *The Rise of Statistical Thinking, 1820-1900*, Princeton University Press, Princeton.

Quetelet, L. A. J. (1848). *Du systéme social et des lois qui le régissent*, Guillaumin, Paris.

Quetelet, L. A. J. (1849). Letters addressed to H.R.H. the Grand Duke of Saxe Coburg and Gotha, on the Theory of Probabilities as applied to the Moral and Political Sciences, Trans. from the 1846 original. London: Charles & Edwin.

Salmon, W. (1980). John Venn's Logic of Chance, in (Hintikka et al., 1980, Vol. II, 125–138.).

Stephen, L. (1950). *The English Utilitarians*, 3 vols., Peter Smith, New York. Reprint of 1900.

Stigler, S. (1986). *The History of Statistics: The Measurement of Uncertainty before 1900*, Harvard University Press, Cambridge.

Venn, J. (1866). *The Logic of Chance*, Macmillan and Co., London. For other editions see (Venn, 1876) and (Venn, 1962 [1888]).

Venn, J. (1876). *The Logic of Chance*, 2nd edn, Macmillan and Co., London.

Venn, J. (1962 [1888]). *The Logic of Chance*, 4 edn, Chelsea, New York. 4th edition identical to the 3rd edition.

Venn, J. (1994). *The Principles of Empirical or Inductive Logic*, Thoemmes Press, Bristol. Reprint of 1889.

Wiener, P. (ed.) (1957). *Values in a Universe of Chance*, Stanford University Press, Stanford.

PART 2

CONTEMPORARY ISSUES IN PROBABILITY THEORY AND STATISTICS

J.B. PARIS

ON THE DISTRIBUTION OF PROBABILITY FUNCTIONS IN THE NATURAL WORLD[1]

ABSTRACT

The purpose of this note is to describe the underlying insights and results obtained by the authors, and others, in a series of papers aimed at modeling the distribution of 'natural' probability functions, more precisely the probability functions on $\{0,1\}^n$ which we encounter naturally in the real world as subjects for statistical inference, by identifying such functions with large, random, sentences of the propositional calculus. We explain how this approach produces a robust parameterised family of priors, J_n, with several of the properties we might have hoped for in this context, for example marginalization, invariance under (weak) renaming, and the possibility, when using the J_n as priors, of non-tautologous universals having non-zero probability.

Keywords: Prior probability, imprecise probability, random sentences, probabilistic reasoning, uncertain reasoning.

1. INTRODUCTION

The purpose of this paper is to describe the current state of an ongoing collaborative research project involving George Wilmers, Paul Watton, Alena Vencovská and the author which started around 1990. Whilst this research can be motivated from several different directions, and indeed those participating did not always even share a common viewpoint on this,[2] one enduring strand was to attempt to model the distribution of probability functions as encountered, and recognised, by us in the real, or natural, world.

The motivation for doing this arose from our contact in the 1980's with the emerging expert system, later to become knowledge based system, technology. Even at that time there were a bewildering multitude of expert system methodologies on offer, bewildering because comparisons between them were difficult.

One reason for this difficulty was because expert systems were frequently customised to suit some very specific applications. A second reason was because the 'correct' answers these expert systems sought to predict were not known in any case.

With this perception of the state of the art at that time we considered if it might not be possible to objectively test expert system methodologies in a similar fashion to the way one might test the output of a computer algorithm or the accuracy of a statistical decision procedure by determining their performance over the range and distribution of their expected or intended applications.

Since we were primarily interested in expert systems which adopted various theories of 'imprecise probabilities' to infer probabilities (or conditional probabilities) in the test material what we primarily sought was a family, or more precisely a distribution, of probability functions similar to that encountered by expert system builders in the course of their business. We shall refer to such probability functions as 'natural' and to their posited distribution as the 'natural distribution'.

Whilst the evaluation of probabilistic expert systems was one particular initial motivation the distribution we consequently sought bears on a related problem which is at the very heart of Bayesian inference, namely the so called 'Prior Problem'. This problem arises, for example, in situations where we have some set K of constraints on some real world, i.e., 'natural', probability function P and we wish to estimate $P(X)$ for certain events X. The standard Bayesian approach here would be to assume that this natural (but, except for K, unknown) probability function P comes from some distribution D of such natural probability functions and take as our estimate of $P(X)$, the expected value of $Q(X)$ when Q is distributed according to D conditioned on K.

The question then, is how to choose the 'prior' D? The base case here is when there are no constraints (i.e., $K = \emptyset$), which, provided we are talking about natural probability functions, is just our problem again. Essentially then what we have here is the age old 'prior problem in the absence of any knowledge' except that we are assuming that we do have one vital piece of knowledge, namely that the unknown probability function P whose values we wish to estimate is natural, does really exist in the real world.

2. UNIVARIATE CASE

In order to appreciate the significance (to our mind) of the assumption that P really exists in the natural world we shall start off by considering the case of univariate P, that is, corresponding to the probability of an event that either does or does not happen.

In this case we can think of P as a natural *random process* which outputs say 1 (as opposed to 0), or true (as opposed to false), with a fixed probability $P(1)$. Notice that P is entirely determined by $P(1)$, equivalently its expected value, $E(P)$.

The question is then, *given only that we know that P corresponds to a natural 0-1 random process how should we assume the possible values of $P(1)$ are distributed, equivalently what prior should we assume?*

In this paper we shall present one approach to this problem. Firstly, however, we should, at least briefly, consider whether this question actually has any meaning at all. For surely, it might be objected, the question very much depends on our interpretation of the world and on the categories we use to define it and as such has no objective sense.

Whilst others may disagree, we do not see this as necessarily invalidating the question. After all *any* modeling of the real world depends on our interpretation of it, the main extension of that situation here is that now 'the real world' also encompasses us ourselves as one of the players.

Indeed we believe one can, in a sense, see such a distribution in existence. Namely consider a very large multivariate natural probability function, say the probabilities of all medical conditions, signs, symptoms etc., and consider all its univariate marginalizations. We can plot these, via their expected 0-1 values, on the interval $[0, 1]$. This gives a sort of natural distribution – and whatever it is it has a shape! [In fact, to our knowledge at least one investigation along roughly these lines has been carried out already, by Egon Pearson in 1925, and we shall later consider his findings in the light of our conclusions.]

A second argument (albeit possibly the million flies argument) that this idea is not entirely meaningless is that the notion of a prior is already accepted, unquestioningly it seems, by the vast majority of practicing statisticians. Yet the prior(s) they apparently have in mind would seem by and large to be far more abstract than the rather concrete suggestions in this paper.

Having briefly addressed the issue of whether or not our question has any meaning we shall now take that as given and comment on the 'standard' Bayesian approach, which is to take as the prior the *uniform distribution* (or somewhat more generally a Dirichlet or beta distribution). As far as this paper is concerned we would criticise this on at least two grounds.

The first is that it seems to be justified by some idea of 'indifference'. This *might* be acceptable if we were discussing the formation of beliefs or subjective probabilities, as reflected in willingness to bet, in some entirely theoretical situation divorced from the real world. However, as an assumption about the real world as we encounter it 'indifference' seems no more to us than an act of faith. Why should it hold? What models do we have that could explain this phenomenon?

Our second criticism concerns its very appropriateness for this role. To illustrate this shortcoming consider the following question: What probability would you give to a particular flower of the plant tormentil having (exactly) four petals?

Surely one's immediate thoughts here would be for a probability close to (though probably not exactly equal to) 0 or 1. And likewise one would surely feel that the answer would be more likely to lay between 0 and $1/8$ than between $1/\pi$ and $1/8 + 1/\pi$, in direct contradiction to uniformity.

This prevalence for natural probabilities to have values close to 0 or 1 seems to be a feature of the world we live in, and fortunately so perhaps, allowing us much of the time to operate on automatic pilot, by default. Indeed the emphasis on the study of defeasible reasoning, that is reasoning with statements of the form

$$\text{if } \theta \text{ then usually } \phi,$$

within the context of intelligent computer systems provides hard currency to support this thesis.

Whilst probabilities close to 0 or 1 do appear especially commonplace there also seems to be, in the natural world, another local concentration, albeit far less significant, around 1/2 (and in turn still lesser ones at 1/4, 3/4 and so on). Indeed, after 'close to 0 or 1' that is surely the next candidate answer in the case of the tormentil's petals. One possible explanation for such a concentration around 1/2 is that there are certain symmetries in the natural world amongst which two way symmetries appear by far the most common.

We have used the example of the tormentil's petals here to argue against the appropriateness of the uniform distribution but can we explain what it is about this example that breaks through the 'indifference'. What does the fact that we were talking about a real plant tell us? More generally, what does simply knowing that the probability function we seek really exists in nature tell, or suggest, to us?

As a central tenet of this paper we would argue it tells us, that, as a 0-1 random process,

The process is very complicated and the randomness is hidden very deeply within it

For example the number of petals on the tormentil flower is not a random event that happens exactly when we looked at is. It is not decided by God throwing a die just before the bud opens. In fact the real randomness is hidden deep, maybe so deep that we could never really isolate that point at which the decision was made. This situation, we would maintain, is typical of the random processes we (with the possible exception of particle physicists) encounter in the natural world.[3]

In view of these considerations (and others) papers (Paris et al., 1991, 1994, 1998, 1999, 2000) represent an attempt to provide a model of such processes and in turn generate a candidate 'natural distribution', that is a distribution on the natural probability distributions, equivalently in this univariate case, natural random 0-1 processes.

3. THE UNIVARIATE MODEL

Our approach was to consider modeling such processes by very large sentences of the propositional calculus, a natural approach given our backgrounds in logic.

Sentences of the propositional calculus yield 0-1 random processes via their truth values (1=true, 0=false) when the truth values of constituent propositional variables are assigned randomly. To give a toy example of what we mean here, if θ is the sentence

$$(\neg p_2 \vee (p_1 \wedge p_3)) \vee \neg p_1$$

and the propositional variables p_1, p_2, p_3 are independently distributed with expected (truth) values 1/2, 1/3, 2/3 respectively then, since θ is true just if one of the disjoint

$\neg p_1$ (expected truth value $1 - 1/2 = 1/2$),
$p_1 \wedge \neg p_2$ (expected truth value $1/2 \cdot (1 - 1/3) = 1/3$),
$p_1 \wedge p_2 \wedge p_3$ (expected truth value $1/2 \cdot 1/3 \cdot 2/3 = 1/9$),

is true the expected truth value of θ is $1/2 + 1/3 + 1/9 = 17/18$.

With this picture in mind an initial attempt might be as follows. Take as models of 0-1 random processes all sentences of the propositional calculus containing exactly n connectives and using propositional variables from $p_1, p_2, ..., p_m$ (for some fixed m, n) where the truth values of the propositional variables are randomly and independently distributed with expected truth values 1/2.

Assuming all such processes equally likely to be encountered this yields a distribution $D^0_{n,m}$ on the expected values of such sentences, i.e., within this modeling a distribution on the natural (univariate) probability functions.

This initial model raises a number of questions. Which connectives should we use in forming our sentences? What values should we give to n and m? What is the justification for assuming that the propositional variables p_i all have the same expected value 1/2?

As far as the connectives is concerned, it has become clear (to us at least) over the course of this research project that this choice is highly influential. Our current view is that in forming these sentences we should allow all *genuinely binary* connectives. By genuinely binary we mean that we exclude the binary connectives which in fact only depend on one argument and those which do not depend on either argument, in other words always return the same truth value (0 or 1) independently of the truth values of the constituent arguments. Actually including the first of these would, in the final analysis, have no effect. On the other hand including the two constant binary connectives would have, within this intended modeling, the rather undesirable effect that a significant proportion of our sentences allegedly modeling random processes would have constant truth value and so would not be random at all!

This observation uncovers a key issue, that of 'pre-selection'. Namely, we are trying to model the distribution of natural 0-1 random process encountered in the real world and it is implicit in this notion that we are limiting ourselves to processes which have been pre-selected as being genuinely, recognizably, random. Our modeling therefore had better not include a significant proportion of models of processes which are not random at all. One way of avoiding this problem is to exclude the constant connectives. (An alternative approach, which ultimately leads to the same results, will be mentioned later).

Finally the choice to limit ourselves to binary, rather than also including ternary etc., connectives, is based on the intuition that, with the possible exception of very high energy physics, interactions in this world ultimately take place between *pairs* of reagents.

Turning now to the choice of n and m, it is clear that to capture the requirement that our processes be very complicated and the randomness (i.e., the propositional variables) hidden very deep both n and m should be very large. And indeed, that to capture this we should take not $D^0_{n,m}$ but instead

$$D^1 = Weak\ limit\ as\ n,m \to \infty,\ D^0_{n,m},$$

a limit which we can show exists.

Having made this improvement we are now in a position to address the third and final question mentioned above, namely, how can we justify assuming that all the p_i have expected truth value $1/2$. For now we could argue that D^1, as our first candidate for a natural distribution, should be an improvement on this simple discrete distribution assumed initially for the p_i and that we should now start again with that distribution. Doing this we can again show that the weak limit D^2 exists. Similarly starting from D^2 we obtain D^3 and so on. In each case we could argue that the resulting distribution was an improvement on its parent, and indeed that the weak limit, J, of these D^i (again it exists), is preferable to all of them.

This limit J is our (current) candidate for the 'natural distribution' in the case of univariate probability functions. Before going on to consider the properties of J we mention a couple of points related to our earlier choice of connectives.

In the earlier paper (Paris et al., 1994) we carried out a similar construction but using only those binary connectives which, with arguments p,q, have the truth tables of

$$p \wedge q,\ p \wedge \neg q,\ \neg p \wedge q,\ \neg p \wedge \neg q,\ p \vee q,\ p \vee \neg q,\ \neg p \vee q,\ \neg p \vee \neg q.$$

[This is equivalent to taking sentences built up from $\{\pm p_i | i = 1, 2, 3, ...\}$ using only the connectives $\wedge, \vee$, which is how it was actually done in (Paris et al., 1994).] In that case the final weak limit of the $D, D^{(2)}, D^{(3)}, D^{(4)}, ...$ again exists, and is the measure U which puts mass $1/2$ on each of the singletons $\{0\}, \{1\}$ and none anywhere else. In other words, in the limit all randomness has disappeared! This 'experiment' certainly indicated to us that we should be rather more catholic in our choice of connectives.

A second point here concerns our rejection of the constant binary connectives. Whilst this may accommodate the pre-selection requirement it may with hindsight seem that there was a preferable method, namely to allow in all the binary connectives and instead to condition the D^i on the open interval $(0, 1)$ (it can be shown that this has non-zero measure with respect to the D^i) and take the weak limit of these conditioned D^i. Pleasingly however this alternative approach also gives the same J in the limit!

4. PROPERTIES OF J

The first property of J worth mentioning is that it is continuous, hence countably additive. A second attractive property of J concerns the question of whether it might not be possible to improve still further on J by starting all over again, but now with the p_i distributed according to J. In fact, as one surely would have suspected, there is nothing more to be gained we just regenerate copies of J all the time. Indeed, no matter what countably additive distribution T we start with, provided $T \neq U$ the analogous process which gave J from the starting distribution D^0 will give J again. [If we do start with U we just get U back again.]

A third pleasing feature of J is that we *know* it in the sense that we can explicitly give all its moments, $E(x^k)$. Precisely, the kth moment is given by,

$$E(x^k) = \frac{2}{5}E(x^k)^2 + \sum_{r=0}^{k} \frac{2}{5}(-1)^r E(x^r)^2 \binom{k}{r}$$
$$+ \sum_{r=0}^{k} \frac{1}{5}E(x^r(1-x)^{k-r})^2 \binom{k}{r}.$$

From these moments we can construct a relative frequency histogram of J using the method of Bernstein polynomials. Figure 1 shows the result of doing this using the first 100 moments.

In terms of our earlier intuitions about the distribution of natural probability functions J certainly seems to have the right shape with conspicuous flick ups approaching 0 and 1 and a discernible bump at 1/2, echoed faintly at 1/4, 3/4/ etc. [A word of warning here however. In (Paris et al., 1994) the measure corresponding to our D^1, whilst in that case continuous, was shown to have infinite derivative at a dense set of points. We currently have no good reason for believing that under the magnifying glass J itself will not prove to be equally spikey.]

Notice also that in J the source randomness has been 'hidden deep' – indeed so deep that it is effectively inaccessible. In this

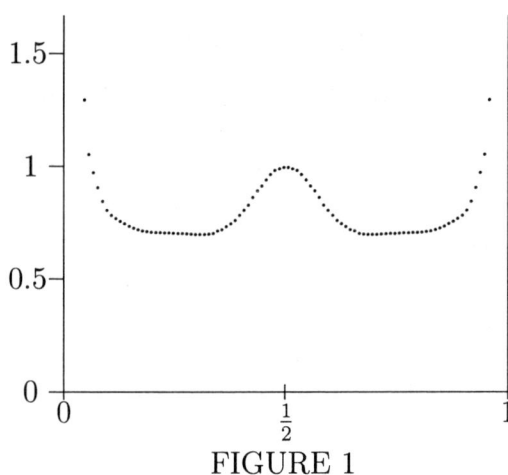

FIGURE 1

model then randomness has no beginning, it is just there, an intrinsic part of the machine!

We conclude this list of 'pleasing properties' of J by mentioning an alternative justification for the choice of J as a natural prior probability distribution. Consider some natural probability function (on SL_1), or equivalently some random 0-1 process in the real world. As we have already argued our experience of such natural probability functions is that this process is ultimately a very complex combination of other processes and that any true randomness is hidden deep down at the microscopic level. Now it seems reasonable to suppose that on closer inspection this process will be seen to be the result of a simple combination of some few other natural processes, where, just as in the initial process, the true randomness is again hidden deep down at the microscopic level. Clearly then, if we suppose that our initial natural probability function was distributed according to a natural prior K then we would seem obliged to afford the same status to these 'few other natural probability functions' whose simple combination yielded our initial function. If we now further agree that the possible 'simple combinations' are just those given by 'genuinely' binary connectives (and each of these are equally likely here) this imposes a fixed point condition on K - whose only solutions are J and U! (See (Paris et al., 1998, Theorem 13).) Discounting U

for reasons already given leads us again then to the distribution J.

To sum up this section, in the univariate case J is our initial candidate for the 'natural distribution'. Of course it depends on a number of assumptions, in particular that sentences provide an adequate model of natural random 0-1 processes. Surely one could think of other ways to model this. However, we would be inclined to conjecture that the general shape of J would be repeated by other modelings provided they satisfied certain reasonable (i.e., appropriate to the situation) constraints, in particular that they were sufficiently complicated and had the real randomness hidden sufficiently deeply.

5. EGON PEARSON'S INVESTIGATIONS

As we mentioned earlier, Egon Pearson (1925) gives account of his investigations concerning the distribution, and other features, of real world probabilities. The data collected by Pearson is quite awe inspiring – in the first instance alone (which is what we will primarily consider here) he took samples of size 35 (actually as 20+15) from some 12448 real world probability functions, selected 'at random'. Examples of such probabilities were:

1. The probability that a man walking down the street will be sporting a bow-tie.
2. The probability that a mare in the Hackney Stud book for 1913 will have foaled in 1908 or earlier.
3. The probability that the first line on a page of *Rewards and Fairies* will contain exactly two verbs.

Pearson's initial interest here was to investigate the assumption of a uniform prior, although he did consider also other priors and common assumptions. Of most interest to us is his (symmetrized) relative frequency histogram of sample means of these 12448 size 35 samples given in figure 2.

The good news in figure 2, as far as our promotion of J is concerned, is that, like the histogram for J, Pearson's histogram is roughly ∪ shaped. Beyond that however serious differences start to appear. Pearson's histogram lacks the flick ups of J at 0

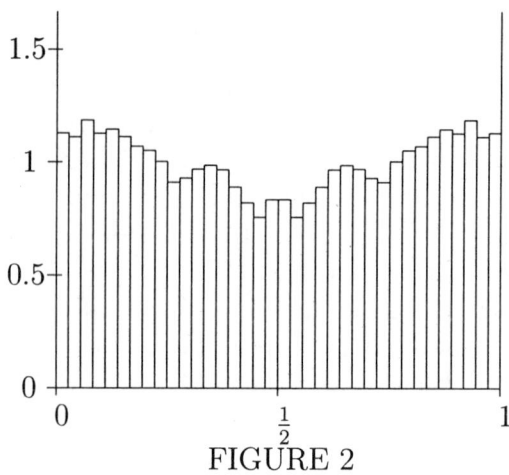

FIGURE 2

and 1, it actually hits its maximum at 0.15 and 0.85 and drops away again beyond these points. Also there is no bump at 1/2. [Indeed on a subsequent smaller set of generally larger samples, the histogram actually dips at 1/2!]

An explanation for this behavior at 0,1, which Pearson partly accepts, is that these probability functions have been, possibly unconsciously, chosen so as to have expected values in the middle, 'visible', ranges. For example that surely influenced the choice of bow-tie, as opposed to say pink carnation, in the first example above. [Interesting it seems from the few examples Pearson specifically cites that in the intervening 75 years several of his 'middling' probabilities have moved close to 0,1.]

An explanation of the behavior at 1/2 is that such expected values frequently arise from symmetry considerations or are otherwise already known to us. Surely in such an investigation there would be a tendency to avoid selecting probability functions whose expected values were already known. [This is not entirely fair, Pearson does also mention using statistics relating to male/female births, deaths.]

In short then his results from our point of view may be argued to be suspect in that these were not a sample of probability functions which *presented themselves*, as they would to, say, a professional statistician in his/her everyday working practice.

There is another sense in which they might not be considered the correct raw material to investigate our J, namely, how far do they go towards being 'natural' in our sense. By and large we would say near enough, however, some sort of a line does need to be drawn here, clearly 'balls in the bag' type probabilities would hardly seem to fall within the scope of our modeling.

A final point to bear in mind here in relation to the shape of Pearson's histogram is that although the number of probability functions Pearson considered is very large, the sample size of 35 for each function is quite small, so that the effects of random statistical variations are by no means negligible and could be expected to produce a considerable smoothing and distortion vis-à-vis any underlying distribution.

It would be interesting to obtain similar data to Pearson's which somehow ensured randomness of the probability functions sampled. Possibly this could be achieved by, for example, systematically collecting statistical data from some long run of a medical journal, though to our knowledge this has not been attempted.

To conclude this section it is interesting to note that whilst Pearson's results do little or nothing to support our candidate, J, the uniform distribution (and indeed Dirichlet distributions in general) fare no better!

6. THE MULTIVARIATE MODEL

We now turn to the more general case of multivariate probability functions, which, of course, is where the interest lies as far as objectively testing expert systems is concerned. The situation here is that we have random variables $q_1, q_2, ..., q_n$ taking values 1,0 and corresponding to positive or negative outcomes of certain events.

A probability function now is determined by the probabilities it assigns to the 2^n exclusive and exhaustive events

$$q_1 = X_1, \; q_2 = X_2, ..., q_n = X_n,$$

where $X_1, X_2, ..., X_n \in \{0, 1\}$, and our question now is what is the distribution of the 'natural' n-ary probability functions (e.g. as encountered by a knowledge engineer).

Again, as in the univariate case, the stock answer as far as Bayesian inference is concerned is the uniform distribution, or possibly, and rather more generally, a Dirichlet prior (see for example (Laplace, 1840; Carnap, 1952; Rosenkrantz, 1981)). And again this raises the question of how this choice can be justified, except by an appeal to 'indifference', which we have (it seems) no good grounds for assuming that nature abides by.

Indeed, in the multivariate case there is a strong argument for rejecting 'indifference' in nature, namely that if we accept indifference, and hence the uniform prior, on n-ary probability functions then the induced distribution on the $(n-1)$-ary marginalizations of these functions will no longer be uniform (i.e., *marginalization* fails for the uniform priors), see, for example, (Lawry, 1994; Lawry and Wilmers, 1994). So we cannot consistently assume indifference in all arities. But, clearly, why should any one arity be more deserving of indifference than any other?!

An alternative approach, which we advocate in this paper, is to extend the modeling of univariate probability functions we introduced early, culminating in the distribution J, to the multivariate case.

As we shall see, to do this we will need to take into account various additional parameters but for the present let us fix our attention on the specific case where we want a distribution on the natural n-ary probability functions which an expert system builder, or knowledge engineer, might in the course of his/her work attempt to estimate. In this case we would anticipate that there would be numerous connections, or correlations, between the various 'events', as represented by $q_1, q_2, ..., q_n$ above. Indeed it is precisely on the presence of such correlations that the knowledge engineer's hopes of ultimate success are built. Without them it would clearly not be possible to make any non-trivial inference about one q_i on the basis of some knowledge about some other q_j's. In order to provide our modeling then we need to provide an explanation as to the nature of these 'connections'.

Clearly in view of our arguments in the univariate case the individual q_i should be distributed according to J. In view of this there is an obvious explanation why different q_i might be interdependent, namely, thought of as large sentences they actually have some subformulae *in common*. Indeed it is hard to see any other mechanism that could generate dependencies.

Exactly how to take the next step and formalise this phenomenon, however, is rather less obvious. One way that we have investigated in some detail is to treat the q_i as compound 'events' depending on common, shared, 'basic', independent features p_j as in figure 3.

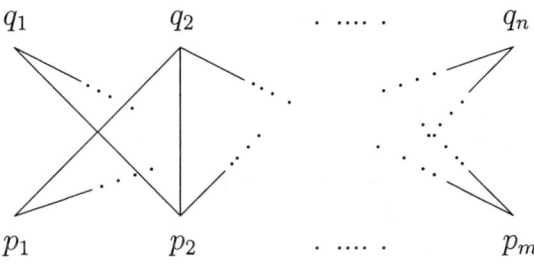

FIGURE 3

As to the nature of these 'compound events', this is where we made our second key assumption. In line with the above motivation we assumed that each q_i was in fact a sentence θ_i built up from the p_j (*without repeats*) using k genuine binary connectives. So our diagram in figure 3 really looks like figure 4.

For example, q_1, q_2 might be

$$q_1 = \theta_1 = (p_2 \wedge \neg p_3) \vee p_4,$$
$$q_2 = \theta_2 = \neg p_2 \wedge (p_1 \vee p_3),$$

and there would be an (apparent) interdependency between q_1 and q_2 since if q_2 holds (i.e., takes value 1) then $\neg p_2$ holds, the first conjunct of θ_1 fails, and hence (apparently) the chance of q_1 holding would be diminished.

Assuming now that the expected values of the p_j, $E(p_j)$, are distributed according to J, and all such θ_i are equally likely gives our (parameterised, by m and k) candidate *natural distribution* J_n of 'natural n-ary probability functions'. (To simplify the notation we suppress specific mention of m, k in J_n.)

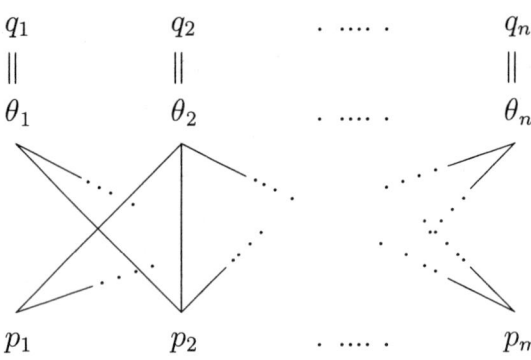

FIGURE 4

Clearly we have made a number of assumptions here which should not pass without comment. Firstly, our idea of using sentences θ_i and the distribution J seems, given our approach in the univariate case, to be almost forced.

Secondly, the need for some parameterisation here appears 'unavoidable' for following reason. If we chose n arbitrary features of the real world then surely they would be independent. On the other hand if we wanted to restrict our attention to the range of n-ary probability functions which present themselves as the raw material of expert system builders then we would want many connections between the features. These differences are achieved in this modeling by adjusting the parameters m, k – and quite appropriately so. Roughly m small increases the strengths of connections between the q_i whilst increasing k has the effect of increasing the number of such connections. [Actually there is a case for arguing that we should further increase the parameters to allow at least weighted mixtures of different m's and k's.]

Thirdly, on the matter of why we placed the restriction on the θ_i that they should contain no repeated propositional variable, p_j, this is entirely natural if we think of the individual q_i themselves large sentences distributed according to J since in that modeling (as described in the univariate case) these sentences will not have repeated subformulae.

Defined in this way the J_n have many attractive properties. For example, the J_n marginalize, which, as we have already seen, is a property not possessed by the conventional uniform distribution(s). Furthermore, when $n = 1$ the J_n are all equal to J, independently of m and k. Clearly this is a very desirable property, it would have been most unfortunate if having argued for J in the univariate case and the J_n in the multivariate case they had disagreed when $n = 1$!

The J_n are also *symmetric* (in the notation of our papers satisfy *weak renaming*) in the sense that they invariant under permuting the q_i, or transposing q_j and $\neg q_j$. (For further details see (Paris et al., 1998).) However, as a measure on probability functions on the events

$$q_1 = X_1, \ q_2 = X_2, \ q_3 = X_3, ..., q_n = X_n,$$

where $X_1, X_2, ..., X_n \in \{0, 1\}$, the J_n are not invariant under arbitrary permutations of these events. A consequence of this is that when using this prior in inductive reasoning Johnson's Sufficientness Postulate (see for example (Johnson, 1932), or (Fine, 1973), or (Paris, 1994)) fails. However, we would assert that in view of the criticisms of this postulate, see for example (Paris et al., 2000), this is a desirable property of the J_n!

As priors for inductive reasoning the J_n have a further pleasing property. Namely, for $n > 1$ using the J_n allows that non-tautologous universals may get non-zero probability. (Again the failure of this for Carnap's continuum of inductive methods is a commonly stated shortcoming of these more conventional priors.) Furthermore, this can happen for 'the right reasons', namely that the J_n allow the possibility that there may be hidden connections between the q_j which force some Boolean combination of them to always hold. In our opinion it is precisely the possibility of such links that fuels our feelings that non-tautologous universals should not, per se, be necessarily assigned zero probability.

Having lauded so many of the J_n's fine properties, we should in fairness note one criticism of the J_n as regards certain applications. Namely, that calculating any but the simplest moments of the J_n appears to be very messy – see for example (Paris et al., 1998).

7. CONCLUSIONS

In this paper we have tentatively proposed a family of 'natural probability distributions' aimed at modeling the distribution of probability functions as we encounter them in the real world. In the course of doing this we made a number of assumptions, in particular:

1. For the unary case, that such 'natural' probability functions correspond to very complicated random 0-1 processes in which the genuine randomness is hidden very deep, and that such processes may be modeled by very large sentences of the propositional calculus.
2. For the n-ary case, that connections between features arise because they corresponded to simple sentences built up from common or overlapping propositional variables.
3. That the 'problem of pre-selection' can be addressed by limiting ourselves to genuinely binary connectives (or allowing all binary connectives but repeatedly conditioning on probabilities in the open interval $(0, 1)$).

How reasonable are these assumptions? One might, of course, argue that the proof of the pudding is in the eating, and that the assumptions receive some credibility from the J_n's having the 'right' properties. However, an appeal to any such defense would seem to be contrary to the spirit of this paper with its pretentions to model and explain rather than simply describe. And judged as an explanation of natural randomness there seems a rather plain criticism of, say, the first two assumptions, namely, that whilst in practice we may see very complicated processes at work we do not transparently 'see' these posited underlying sentences. How can we answer that?

One response might be to argue that our current possible failure to see things in this way should not necessarily be taken to mean that they cannot be so seen. But be that as it may, we would still contend that this modeling may, nevertheless, capture the truth *by analogy*. That viewpoint would, of course, look more reasonable if similar conclusions were also found to follow by using various other models of 'complicated processes'. If so it would be an interesting result, analogous in a way to the law of large numbers, since it would say that provided a process was complicated enough and the true randomness hidden deeply enough, then it will, by its very nature, acquire certain common, or 'universal', characteristics. But for the present, this, of course, is pure speculation.

ACKNOWLEDGEMENTS

I would again like to explicitly mention my collaborators, George Wilmers, Paul Watton, Alena Vencovská, in the research project on which this paper is based. I would also like to extend my thanks to Costas Dimitracopoulos for his encouragement and practical support and to Peter Walley, Peter Wakker, Terence Fine and the various referees of earlier papers on this topic for their valuable comments and suggestions.

POSTSCRIPT

With reference to the derivation of natural distributions suggested on page 127, in a recent paper Jon Williamson (n.d.) includes a frequency distribution of the (marginal) probabilities of source nodes from a selection of 34 Bayesian network databases from the Machine Learning Repository, (Blake and Merz, n.d.). His graph shows some clear similarities with J, in particular sharp flick-ups at 0 and 1 and a substantial peak around $1/2$. The most obvious difference is that probabilities in his intervals $[0.2, 0.35]$ and $[0.65, 0.8]$ are significantly under-represented compared with the predictions of J.

Department of Mathematics
University of Manchester
England

NOTES

[1]This research was partially supported by a joint British Council/University of Athens Research Grant.

[2]For that reason 'we' in the text should be taken to mean possibly only a subset of the responsible investigators, although including at least the author!

[3]So we are not talking 'balls in urns' here, or similar mental constructions, a model which seems to figure remarkably large in the subject given its artificiality.

REFERENCES

Blake, C. and Merz, C. (n.d.). *UCI Repository of machine learning databases*, http://www.ics.uci.edu/~mlearm/MLRepository.html, Irvine, CA: University of California, Department of Information and Computer Science.

Carnap, R. (1952). *The Continuum of Inductive Methods*, University of Chicago Press.

de Cooman et al, G. (ed.) (1999). *Proceedings of the First International Symposium on Imprecise Probabilities and Their Applications (ISIPTA-99)*, Ghent, Belgium.

de Glas, M. and Gabbay, D. (eds) (1991). *Proceedings of the First World Conference on the Fundamentals of AI*, Paris.

Fine, T. L. (1973). *Theories of Probability*, Academic Press, New York.

Johnson, W. E. (1932). Probability: The deductive and inductive problems, *Mind* **49**: 409–423.

Laplace, P. S. (1840). *Essai Philosophique sur les Probabilités*, Bachelier, Paris.

Lawry, J. (1994). *Natural Distributions in Inexact Reasoning*, PhD thesis, Manchester University, Manchester.

Lawry, J. and Wilmers, G. M. (1994). *An axiomatic approach to systems of prior distributions*, in (Masuch and Polos, 1994).

Masuch, M. and Polos, L. (eds) (1994). *Knowledge Representation and Reasoning under Uncertainty*, Lecture Notes in Computer Science, Springer-Verlag, Berlin.

Paris, J. (1994). *The Uncertain Reasoner's Companion – A Mathematical Perspective*, Cambridge University Press.

Paris, J., Dimitracopoulos, C., Vencovská, A. and Wilmers, G. (1998). A multivariate natural prior probability distribution based on the propositional calculus, *Technical Report of the Manchester Centre for Pure Mathematics*, Department of Mathematics, University of Manchester, Manchester.

Paris, J., Vencovská, A. and Wilmers, G. (1994). A Natural Prior Probability Distribution Derived from the Propositional Calculus, *Annals of Pure and Applied Logic* **70**: 243–285.

Paris, J., Vencovská, A. and Wilmers, G. M. (1991). *A Note on Objective Inductive Inference*, in (de Glas and Gabbay, 1991).

Paris, J., Watton, P. and Wilmers, G. (1999). *On the distribution of natural probability functions*, in (de Cooman et al, 1999, 302–311).

Paris, J., Watton, P. and Wilmers, G. (2000). On the structure of probability functions in the natural world, *Uncertainty, Fuzziness and Knowledge-Based Systems* **8**: 311–329.

Pearson, E. (1925). Bayes theorem examined in the light of experimental sampling, *Biometrica* **17**: 388–442.

Rosenkrantz, R. (1981). *Foundations and Applications of Inductive Probability*, Ridgeview Publ. Co.

Williamson, J. (n.d.). *Random probability functions*, philosophy.ai report pai_jw_00_e, http://www.kcl.ac.uk/philosophy.ai.

GLENN SHAFER

NATURE'S POSSIBILITIES AND EXPECTATIONS[1]

A flexible and commonsensical theory of causality can be based on the idea of Nature's evolving predictions. Nature witnesses the unfolding of events at levels of detail finer than that of any actual witness, and she makes predictions of future events that are never falsified. Although we seldom see events as Nature sees them, we often conjecture about Nature's predictions from regularities we do see, and we sometimes build these conjectures into our own reasoning and prediction. Mathematically, these conjectures concern possibilities and expectations in Nature's event tree.

The regularities we witness in the world are often called "Nature's laws." Who is Nature? She is only a metaphor, but this metaphor can take different shapes in our imagination. Sometimes we think of Nature merely as a witness; she witnesses what has happened so far, and using her laws, she can predict something about what will happen next. When her laws are categorical, she predicts the future with certainty, when her laws are only probabilistic, she gives odds for what will happen. On other occasions, we fancy Nature as an actor: She makes things happen according to her laws. When the laws are probabilistic, she is the one who rolls the dice and then enforces the outcome.

In science, the metaphor is kept at a distance. It may inspire theory, but it has no place in the formulation of hypotheses and their empirical verification. So the shape of the metaphor does not matter; anything goes. In the philosophy of science, on the other hand, the difference between a passive and active Nature appears to be significant, especially when we try to understand causality and the closely related concept of objective probability. Making Nature an actor gives us a vivid sense of the reality of these ideas and preserves a good deal of their mystery. But in a philosophy of science in which Nature appears merely as a witness and predictor, the mystery (and excitement) of causality and objective probability seem to drain away; there is nothing there but prediction.

In *The Art of Causal Conjecture* (Shafer, 1996a), I argued for a passive Nature and a predictive understanding of causality. Causal regularities, I argued, are regularities in the temporal unfolding of events. Objective probabilities arise when these regularities are less than uniform. In this article, I pick up this theme and try to place it more clearly in its historical and philosophical context.

I begin by reviewing the flexibility of event trees as a representation of causal structure. The concept of an event tree for Nature leaves us free to assert or deny that Nature witnesses causal regularities of a given form in a given domain. Competing mathematical representations of causal structure, inasmuch as they restrict us to particular kinds of event trees, do not have the same flexibility. A stochastic process, for example, is equivalent to an event tree that is supplied with a precise time scale and a full probability specification. We will look at some simple examples that illustrate how restrictive this is.

1. DYNAMIC REGULARITY IN NATURE

Causal relations are dynamic regularities—regularities Nature witnesses and predicts as events unfold. The branches in Nature's event tree represent the possibilities Nature foresees for the step-by-step evolution of her knowledge, and probabilities on these branches express her limited ability to predict the direction the evolution will take. Although they are subjective probabilities for Nature, we may think about them in much the way statisticians are accustomed to thinking about objective probabilities. They are based on Nature's past experience, and they will be played out in Nature's future experience. If Nature posted her probabilities as betting offers, she would at least break even, approximately, against any opponent. In particular, the frequency with which events happen in a sequence of trials singled out by an opponent would approximate Nature's average probability for those events.

Figure 1 shows how Nature might predict the behavior of Rick, a youngster at home alone on a summer afternoon. Since it has a probability on each branch, the event tree in this figure can also be called a probability tree. The probabilities are Nature's predictions for what will happen in each situation. At the beginning of the afternoon, Nature does not know for sure whether Rick will attend to his bicycle tire. But based on her experience observing him and similar youngsters in similar situations, she gives odds of 4 to 1 that he will.

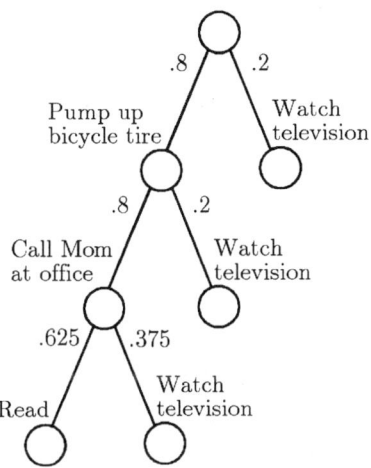

FIGURE 1. Nature's event tree for Rick's choices on a summer afternoon. The numbers are Nature's predictions (probabilities) for what will happen in each situation (circle).

In order to grasp fully the idea of Nature's evolving predictions, we must understand the refinement and simplification of event and probability trees (see (Shafer, 1996a, Ch.13), and (Shafer, 1998)). Nature's tree is presumably exceedingly complex, involving details that go far beyond our own current perceptions and preoccupations. Any tree we might draw is necessarily a simplification. But simplification is not necessarily falsification. Two probability trees, one more detailed than another, can both be accurate representations of Nature's limited ability to predict. Figure 2 illustrates the point; there the simpler tree on the right is consistent with the more refined tree on the left;

both give the same initial probability for Rick's eventually reading (.8 · .8 · .625 = .4) and for his eventually watching television (.2 + .8 · .2 + .8 · .8 · 375 = 6). We may call both "Nature's tree."

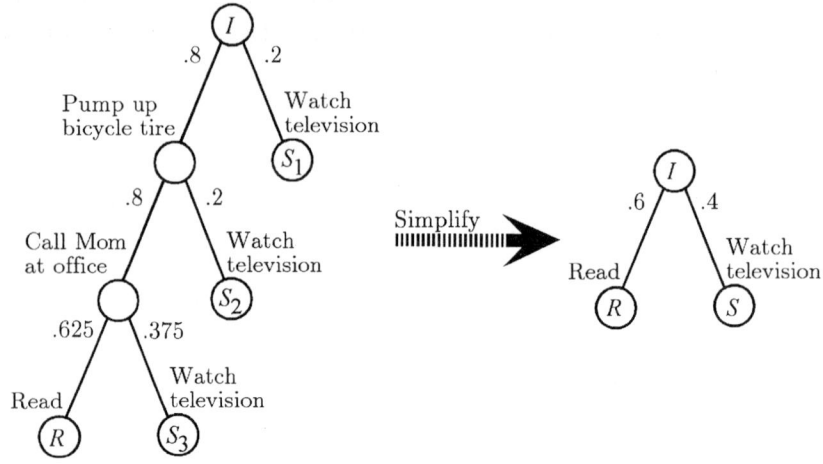

FIGURE 2. A simplification of figure 1. The node S in the simplification has the same meaning as the collection $\{S_1, S_2, S_3\}$ of nodes in the refinement. Nature is in S precisely when she is either in S_1, in S_2, or in S_3.

In general, a node in an event tree represents an instantaneous event (the event that Rick starts watching television, for example) or, equivalently, a situation-the situation in which the instantaneous event has just occurred. Each instantaneous event in a valid simplification must also be represented in the tree it simplifies, possibly as the disjunction of several divergent instantaneous events. (Rick's starting to watch television is shown as a single node in the simplified tree but as three distinct nodes in the refined tree.)

We need not suppose that the causal regularities Nature witnesses and predicts can always be expressed by probabilities, and consequently the refinement of a probability tree for Nature may produce an event tree for which we cannot put probabilities on every branch. For example, the tree on the right in figure 2 might be valid even though there is no refinement of the kind illustrated

NATURE'S POSSIBILITIES AND EXPECTATIONS

on the left. In other words, Nature might assign a 60% probability to Rick eventually watching television without being able to make even probabilistic predictions about whether he will first pump up his bicycle tire or call his mother. This possibility is elaborated in figure 3.

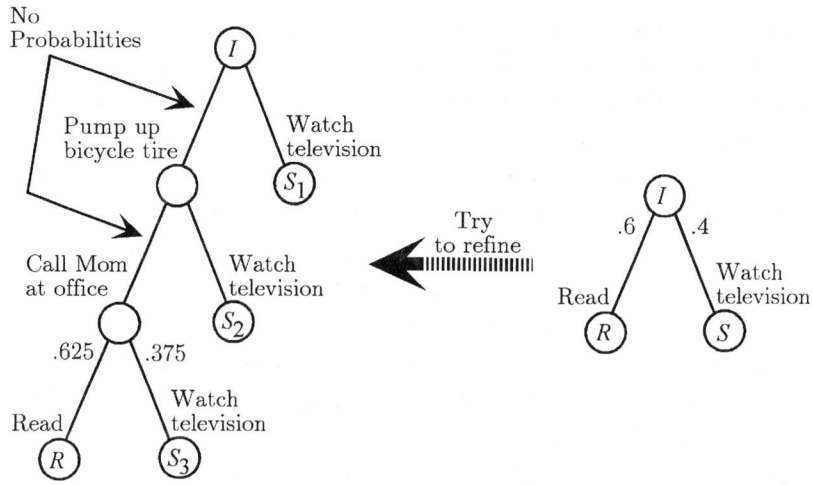

FIGURE 3. Here we suppose that the probability tree on the right is valid for Nature; in situations like I, Nature finds that Rick reads 40% of the time and watches television 60% of the time. These frequencies are stable through time. But this stability of frequencies is missing when Nature tries to predict whether Rick will pump up his bicycle tire or call his mother before reading or watching television. In situations like these, Nature sees frequencies to vary over time and is unable to explain the variation.

2. NATURE AS AN IDEALIZATION

The adjective "ideal" is often applied to simplifications of reality, as when we neglect friction in physics or the width of a line in geometry. When we simplify reality in this sense, we sacrifice precision in order to make a theory easier to understand and easier to use. The simplification is not precisely correct as a description of any real situation, although we can say that it represents the limit of correct descriptions of a sequence of real situations, in which the complications the simplification ignores are successively less important quantitatively.

The concept of Nature is also a limiting idealization, but in almost an opposite sense. There are regularities in the world that actual witnesses can and do see and that actual scientists can and do predict. Nature is the imagined limit as we consider witnesses and scientists who can see and predict more and more. This limit is a simplification of reality inasmuch as it neglects limitations in the knowledge and computational capacity of real scientists. But the ideal structure of prediction that is imagined is not a simplification. On the contrary, it is the indefinitely complicated limit of a sequence of increasingly complicated structures of prediction.

This concept of Nature, a limit of actual and potential witnesses, provides a way of understanding the thought of Jacob Bernoulli, the seventeenth-century Swiss scholar who first made Pascal and Huygens's theory of games of chance into a theory of probability (Shafer, 1996b).[2] Bernoulli did not talk of Nature as a witness, but this idea allows us to complete and make fully coherent Bernoulli's concept of probability. Like every respectable scholar of his time, Bernoulli rejected the idea that events can be determined by Blind Chance. There is no room for chance to determine events, because all things have been foreseen and determined by God. We have probabilities only because, unlike God, we do not know what will happen. All probability is subjective. Yet Bernoulli's subjective probability is a far cry from the subjective probability of twentieth-century Bayesians. We do not learn it by introspection. Often we can learn it only approximately, by long observations of frequencies in the world, and in this sense it is decidedly objective.

In order to see the coherence in Bernoulli's thinking, we must imagine an ideal level of knowledge, a level intermediate between us and God. Bernoulli's probabilities are the subjective probabilities of a witness at this intermediate level. They are subjective inasmuch as they are relative to the knowledge of this witness.

But they are objective inasmuch as the witness is ideal. These probabilities are validated by what actually happens in the world, allowing actual witnesses to approach the position of the ideal witness through their experience.

Why call the ideal witness Nature? I make this claim on the word "Nature" in order to make clear my agreement with Bernoulli's rejection of determination by chance and my disagreement with those who use Nature as a modern and approving synonym for Blind Chance. In the twentieth century, rejection of Blind Chance no longer goes without saying. The idea of determination by chance is very alive in our culture and accepted in much of our scientific literature. Everyone who reads applied probability and statistics understands that phrases such as "stochastic mechanism" and "random process" are meant to evoke determination by chance. Classical (i.e., non-Bayesian) statisticians often speak of the random determination of steps in a stochastic process by Nature. Nature has become an actor, who rolls her dice and uses the outcome to decide what to do. This is pleasantly anthropomorphic, but in my view empirically empty. Once we have said that Nature cannot predict what will happen any more than if she were rolling dice, nothing is added by pretending that she does roll dice and then acts on the outcome. In order to stay in the realm of the empirically meaningful, we should content ourselves with the idea of Nature as witness and predictor, for in this role Nature is the idealized limit of actual or potential scientists, and we can cash out statements about what Nature can or cannot predict in terms of our own achievements and eventual ambitions with respect to prediction.

In contemporary scientific discourse, "nature" represents quite broadly the ground intermediate between the human witness and bare reality. The laws of nature, which we may sometimes perceive at least through the glass darkly, are laws that reality follows, regardless of how we imagine reality to be determined. By claiming "Nature" as the name of my ideal scientist, I stake a claim on this intermediate ground, a claim to be upheld equally against incursions by those who would exaggerate the role of metaphysical suppositions about the determination of reality and those who would exaggerate the role of the actual solitary witness. I mean to reject both the classical statistician's Blind Chance and the Bayesian statistician's insistence on using probability only to describe opinions of actual witnesses. By analyzing causality in terms of "Nature's predictions," I acknowledge the objective nature of causality – its independence of the limitations of specific witnesses – while at the same time rejecting the notion that it depends on some untestable metaphysics of determination.

3. TOWARDS AN INTELLECTUAL HISTORY OF NATURE AS IDEAL WITNESS

I have presented the idea of Nature as ideal witness as my own elaboration of the thinking of Jacob Bernoulli. This is an appropriate acceptance of responsibility; I do not want to condition my adoption of the idea on the claim that particular historical figures would agree with me. But I must also acknowledge other predecessors. In fact, the idea of an ideal witness (if not the name "Nature" for her) has a long history. Many of Jacob Bernoulli's most thoughtful successors, including Antoine Augustin Cournot (1801–1877), Charles Sanders Peirce (1839–1914), and Frank Plumpton Ramsey (1903–1930), resorted to an ideal witness, conceived of as a limit of actual and potential witnesses, in order to explain objective probability.

Cournot, a prolific French mathematician, economist, and philosopher, deserves to head this list, for he developed the idea of an ideal witness in a whole series of treatises (Cournot, 1843, 1851, 1861, 1875).[3] His name for what I call Nature was "l'intelligence supérieure."[4]

Cournot borrowed the idea of a superior intelligence directly from the French mathematician Laplace (born Pierre Simon, 1749–1827), who had used it to explain not probability but determinism. As Laplace explained in 1776,[5] a sufficiently superior intelligence, one capable of apprehending all the details of the present state of the world, could predict the future fully and perfectly from the present using a small number of laws. Laplace chose to emphasize a fictional superior intelligence in his formulation of determinism in order to drive home the point that we

NATURE'S POSSIBILITIES AND EXPECTATIONS 155

humans are in a less exalted position. We must rely on probability and on the mathematical theory of probability, a theory in which, as it happens, Laplace was already the unsurpassed master.[6]

Although Laplace found it convenient to proclaim Nature deterministic and probability subjective (this served to place his work on astronomy at the pinnacle of science while at the same time glorifying his work on probability as the basis of human reasoning), he was a careless philosopher, and in practice his probabilities seemed objective at least as often as subjective. In order to remedy this incoherence in his great mathematical predecessor, Cournot gave his own twist to the idea of a superior intelligence. According to Cournot such an intelligence would have capacities analogous to those of humans but far more powerful—she would be neither God nor man, but "would have a place only in the theological World of the good and bad angels" (Cournot, 1875, 70).[7] And she would differ from us not by dispensing with probabilities but by getting them right (Cournot, 1843, 60). Hers would be the objective probabilities.

Cournot accepted determinism at least in part; he agreed that many individual processes could be predicted in the fashion that Laplace imagined. But he believed that such processes interact in fortuitous ways, foreseen by God (the sovereign intelligence) but not by any intelligence, even superior and theoretical, whose capacity of reason is analogous to our own. Such fortuitous interactions (as when, to cite Cournot's favorite example, a roof tile, following its determinate course, hits the head of the philosopher, heading on his own independent determinate course towards a mailbox) give reality to the idea of chance or objective probability.

Cournot's idea of the intersection of independent causal lines was convincing to hardly any of his nineteenth-century readers. Among French philosophers, it was discussed only to be rejected.[8] It certainly found no place in the doctrines of his British contemporaries Leslie Ellis and John Venn, who based their own version of objective probability on a less subtle equation of probability with frequency. But his idea of a superior intelligence who has objective probabilities—or least his idea of objective probabilities as probabilities scientists would approach in the limit after indefinite investigation, finds echoes in the work of many later writers, including Charles Sanders Peirce in the nineteenth century and Frank Plumpton Ramsey in the twentieth.[9]

We find further echoes in recent work by analytic philosophers. D.H. Mellor, after discussing the idea of objective probability for many years, writes recently, "Maybe the All-Seeing should have no degrees of belief other than 1 and 0, but he can still know the world to be such that we should" (Mellor, 1994, 253). Yet more recently, in Michael Woods' posthumous work on conditionals (1997, 83-84), we find the conclusion that objective probability is subjective probability from "an ideal epistemic standpoint." Woods' ideal epistemic standpoint is my Nature.

In sum, the idea that objective probabilities are the probabilities of an ideal witness has a long history. Aside from my adoption of "Nature" as the name of the ideal witness, the main innovation in my work is to emphasize the evolution of her probabilities and to locate the meaning of causality in this evolution. Whether the course of the world is predetermined by God, left to Blind Chance, or simply chaotic are metaphysical questions with no bearing on the truths of causality. Causality is an idea with empirical meaning, rooted the possibility of prediction and its limits.

4. THE INADEQUACY OF STOCHASTIC PROCESSES

Causal understanding in terms of probability trees does not necessarily reduce to causal understanding in terms of stochastic processes. A stochastic process, interpreted causally, is equivalent to a special kind of probability tree for Nature—one in which the instantaneous events represented by situations are all labeled with precise physical times. In general, a situation in a probability tree for Nature cannot be labeled with a precise physical time unless it is refined, as in figure 4.

One might think, at first blush, that causal regularities expressed in a probability tree can always be refined to more detailed regularities expressed in terms of events at specified times as in figure 4. Such refinement is indeed always possible in a purely

NATURE'S POSSIBILITIES AND EXPECTATIONS

mathematical sense, but we have no guarantee that it will be valid in the Nature's experience. As we saw in figure 3, Nature's ability to make probabilistic predictions may disappear when she refines her event tree. Experience teaches us that regularity can dissolve into irregularity when we insist on making our questions too precise, and this lesson applies in particular when the desired precision concerns the timing of cause and effect. It applies to the natural sciences (where the timing of events may depend delicately on initial conditions) as well as the social sciences. Since Nature represents a limit of capacities of actual scientists, this lesson applies to her as well.

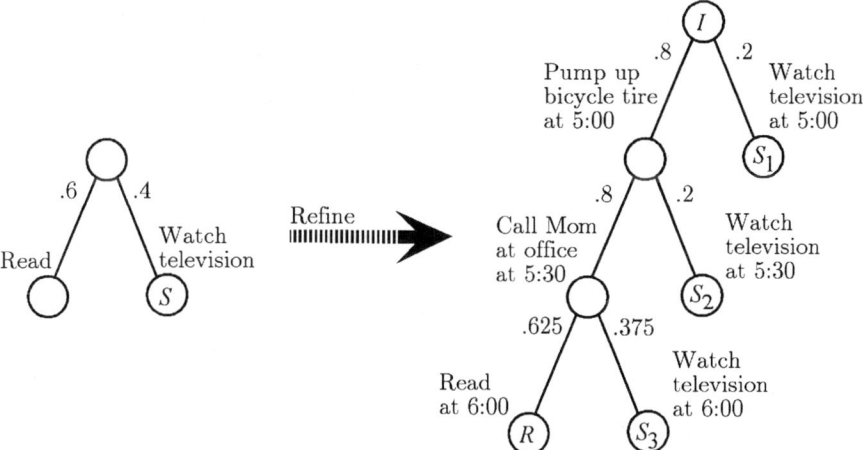

FIGURE 4. Here the more refined tree breaks the instantaneous event S into more specific events that have precise physical times. The more refined tree represents a discrete stochastic process, whose successive steps happen at fixed physical times no matter how events unfold. The less refined tree, on the other hand, cannot be interpreted in this way.

5. A FRAMEWORK FOR CAUSAL DEBATE

By recognizing that causal structures do not always extend to equally crisp deeper causal structures, we can keep causal debates within the bounds of common sense. Suppose the probability tree in figure 5 is Nature's tree for a certain society. This means that no conceivable scientist, no matter how much she witnesses, can improve on its predictions of the sex, education, or income of a particular person in that society. At the point, for example, where a girl is conceived, no scientist observing the conception can do better than give 50–50 odds on whether she will get 8 years of schooling or 12. This is how the society works; as Nancy Cartwright (1996) puts it, the socio-economic machine operates so that we get this result. When people differ about the desirability of the results given by such a machine, they are likely to want more causal information. How is the machine made to work in this way, and what is the range of possibilities for how it might work differently? Though important, these questions do not necessarily have crisp answers, and meaningful debate is possible only when we acknowledge this.

Consider the determination of a woman's level of education. Perhaps Nature can say something about this. Perhaps when Nature witnesses certain events in a woman's childhood she changes her predictions about how much schooling the woman will receive. figure 6 gives an example, in which we suppose that the experience of being a girl scout encourages further schooling. But it is not guaranteed that Nature will witness such regularities. Perhaps the proportion of girls becoming scouts and the schooling received by scouts and non-scouts varies so unpredictably that Nature cannot make probabilistic predictions, as indicated in figure 7. Perhaps there are no signals that can help Nature predict in advance the amount of schooling a girl will get.

If figure 6 is a correct description of how the society operates, then the question of further refinement immediately arises. If steps are taken to get more young girls into the scouts, will more women complete 12 years of schooling? If, for example, the mothers in the Parent-Teacher Organization succeed in enrolling 90% of girls in the scouts, will the proportions of scouts and non-scouts finishing 12 years of schooling remain unchanged, so that the total proportion of women finishing 12 years increases from 50% to 70%, as in the tree on the left of figure 8? And will most

NATURE'S POSSIBILITIES AND EXPECTATIONS   159

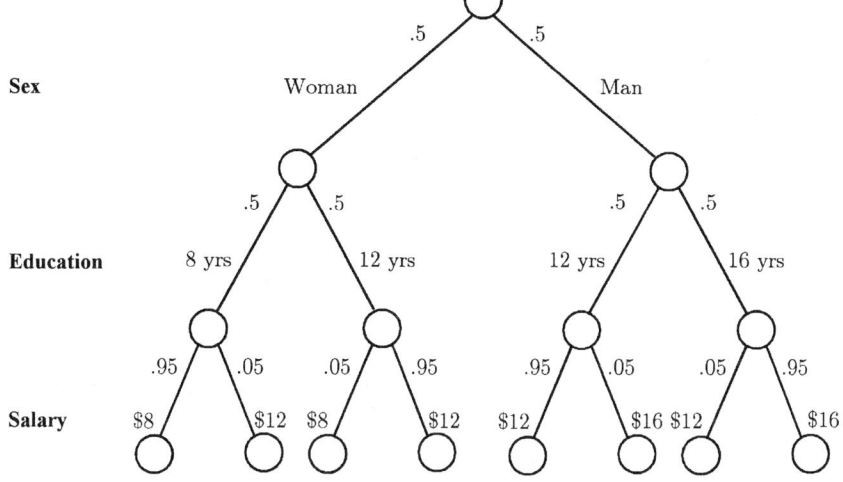

FIGURE 5. The norms of an imaginary discriminatory society. This society educates men more than women, but there is some overlap. People are usually paid in proportion to their education, but employers may deviate from proportionality for an exceptionally capable or hapless employee, provided they stay within the range of pay customary for the employee's sex.

of these better-educated women get better paying jobs, as also indicated in that tree? Or will the society perhaps persist in limiting the proportion of women with 12 years of schooling to 50%, as in the tree on the right in figure 8?

Common sense says that these questions do not necessarily have determinate answers. Whether the new scouts clamor for more schooling, whether more schools are built and more teachers hired-this may depend on how well the new scouts are mentored, who control school finances, and countless other contingencies. Or it may simply be unpredictable. When Nature sees the circumstances, she may or may not be able to make some predictions.

It is sometimes helpful to think in terms of the breadth of a particular causal claim. We begin by supposing that figure 6 applies to a particular society, situated in a certain time and place.

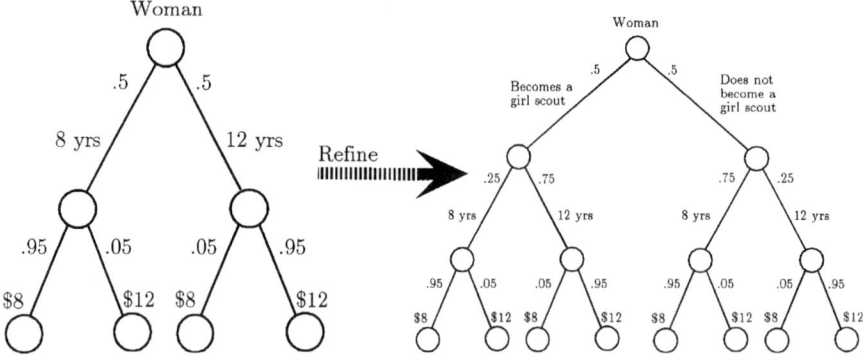

FIGURE 6. Some detail about how the educational level of a woman is determined. Notice that the refinement agrees with all the causal assertions in the original tree. When a woman leaves school after eight years, Nature gives her a 5% chance of earning $12. How she decided to leave school does not matter.

Then we ask how much more widely some of its aspects apply. To what extent does the particular relation among sex, scouting, and education that Nature witnesses in this society apply to slightly different societies at slightly different times? We can expect only very nuanced answers to such questions. Causal relations in a particular society can often be extrapolated only to a limited extent.

Regularities witnessed by Nature, when known to individuals, can be used by those individuals as a guide to action. In a society where figure 6 holds, a mother who wants her daughter to have more schooling will be wise to encourage the daughter to become a girl scout. But such guides to action become less reliable as we move outside the circumstances where we know Nature witnesses the regularity. When all the parents of girls change their behavior, the society itself has changed, and what causal regularities we or Nature will then witness is a new question Lieberson (1985).

NATURE'S POSSIBILITIES AND EXPECTATIONS 161

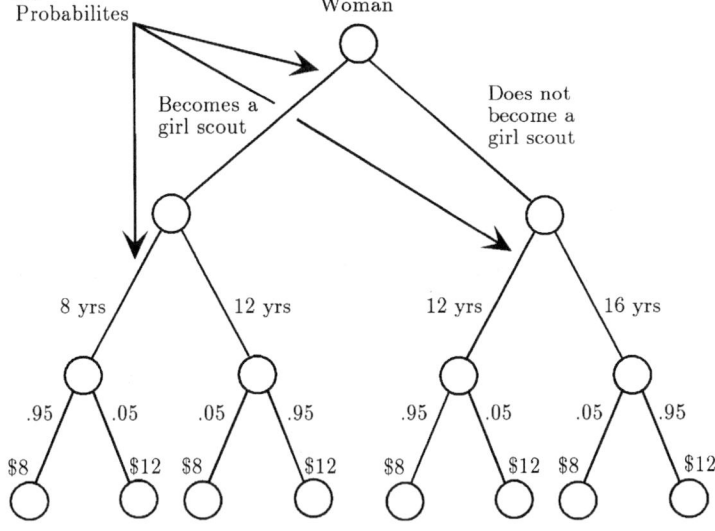

FIGURE 7. In this version of the story, which remains consistent with figure 5, Nature does not witness any stable pattern in the proportion of girls who become girl scouts or in the proportion of girl scouts and non-girl scouts who finish 12 years of schooling.

6. DETERMINISM AND FREE WILL WITHIN NATURE'S EVENT TREE

As Laplace's formulation makes clear, determinism is a hypothesis about predictability. One can believe that all things were determined before the beginning of time, perhaps by God, without being a determinist in Laplace's sense. Determinism goes further; it says that a sufficiently perceptive and well-informed witness can predict the course of events. There are general laws, sufficiently simple that this ideal witness can use them, together with initial conditions, to predict the future fully and exactly. Thus determinism amounts to a special hypothesis about Nature's event tree, the hypothesis that Nature can predict every step in her tree with probability one. According to this hypothesis, the tree does not really branch; it is merely a long chain of inevitable steps.

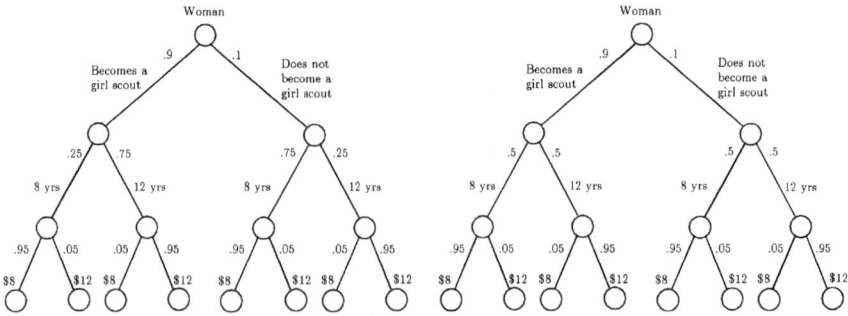

FIGURE 8. In the story on the left, the larger proportion of girl scouts leads to a larger proportion $(.9 \cdot .75 + .1 \cdot .25 = .7)$ of women finishing 12 years of schooling. In the story on the right it does not.

Conversely, when we reject determinism and suppose that Nature's event tree does branch, we are not necessarily rejecting the hypothesis that God knows everything in advance. We are merely rejecting the hypothesis that any human-like intelligence could predict everything in advance.

It is generally agreed that determinism has been refuted by the success of quantum mechanics. This refutation may be inconclusive at the macroscopic level where most discussions of causality are located, but here, too, there is now widespread sentiment against determinism. The conception of causality advanced in this paper can only reinforce this sentiment, for the predictions we can make in the social, biological, and practical sciences are so far from categorical that it seems plausible and reasonable to hypothesize at most probabilistic knowledge for an ideal witness who represents the limit of what we might achieve.

By freeing us from any commitment to determinism, the idea of Nature's event tree allows us to understand causality in the most natural way. As S.N. Bernstein (1932) pointed out, causality is inherently in conflict with determinism, for causality requires possibility: There must be more than one way that an event can come out if how it comes out is to make a difference in how something else comes out.

The most debated aspect of determinism's suppression of causality is, of course, its suppression of free will. It is difficult to see how there can be freedom without possibility. In contrast, the conception of causality represented by a branching event tree for Nature poses no problem for free will. While remaining entirely agnostic about whether predestination by God or later determination by Blind Chance somehow renders the freedom of the individual illusory, we may suppose that Nature, at any rate, participates fully in the illusion. From Nature's point of view, an individual is free to perform an act precisely when Nature cannot predict whether he will perform it or not.

Rutgers Graduate School of Management
Newark, New Jersey
USA

NOTES

[1] Research for this article was supported by NSF Grant SES-9819116. The research has also benefited from conversations with colleagues too numerous to list.

[2] See (Bernoulli, 1713) for a reference to Bernoulli.

[3] For a recent philosophical appreciation of Cournot's ideas, see Martin (1996).

[4] I am endebted to Bernard Bru for directing my attention to Cournot's, which I had not studied for over twenty years.

[5] This is the date when Laplace first declared his determinism in print. It later played a central role in his *Essai philosophique sur les probabilités*, in its many editions (de Laplace, 1814) to (de Laplace, 1825), see (Bru, 1986, 251–289).

[6] Both the originality and influence of Laplace's determinism are often exaggerated. His conception of determinism was not at all original, and it was at odds with the view of many of the contributors to probability who preceded him in the eighteenth century or followed him in the nineteenth. Many of them shared with Jacob Bernoulli the more pious view that the future is up to God, who is not obliged to determine it in a manner explicable to a human-like intelligence, no matter how superior. Gottfried Wilhelm Leibniz (1646–1716), for example, believed that only finite things can be predicted, whereas infinite things remain contingent, being known to God by vision rather than

by demonstration (Parmentier, 1995, 28). And the inventors of statistical physics, especially the British scientists James Clerk Maxwell (1831–1879) and William Thomson (1824–1907), had views closer to Leibniz than Laplace (Smith and Wise, 1989, 430, 632). Nor was Laplace's subjective probability popular with the determinists of the nineteenth century, even in France. The French determinists most influential in science, Auguste Comte (1789–1857) in physics and Claude Bernard (1813–1878) in medicine and biology, were sharply hostile towards probability.

[7]My translation.

[8]It should be added that in recent decades Cournot has received much greater notice from French philosophers of science. His books were all republished by Hachette during the 1970s and 1980s.

[9]In 1928, Ramsey wrote, "Chances are degrees of belief within a certain system of beliefs and degrees of belief; not those of any actual person, but in a simplified system to which those of actual people, especially the speaker, approximate" (Ramsey, 1990, 104ff). Later, he adds, "We do, however, believe that the system is uniquely determined and that long enough investigation will lead us all to it. This is Peirce's notion of truth as what everyone will believe in the end; it does not apply to the truthful statement of matters of fact, but to the scientific system" (Ramsey, 1990, 161). See also (Sahlin, 1990, 115).

REFERENCES

Bernoulli, J. (1713). *Ars Conjectandi*, Basel.

Bernstein, S. N. (1932). *Sur les liaisons entre les grandeurs alé atoires*, in (Saxer, 1932, 288–309).

Bru, B. (1986). Postface to (de Laplace, 1986, 245–303).

Cartwright, N. (1996). *What is a causal structure?*, in (McKim and Turner, 1996).

Cournot, A. (1843). *Exposition de la théorie des chances et des probabilités*, Hachette, Paris. Reprinted as Volume I (B. Bru, editor) of the Oeuvres Complètes, 1984, Paris: J. Vrin.

Cournot, A. (1851). *Essai sur les fondements de nos connaissances et sur les charactères de la critique philosophique*, Hachette, Paris. Reprinted as Volume II (J. C. Pariente, editor) of the Oeuvres Complètes, 1975, Paris: J. Vrin.

Cournot, A. (1861). *Traité de l'enchaînement des idees fondamentales dans les sciences et dans l'histoire*, Hachette, Paris. Reprinted as Volume III (N. Bruyère, editor) of the Oeuvres Complètes, 1981, Paris: J. Vrin.

Cournot, A. (1875). *Matérialisme, vitalisme, rationalisme. Etudes sur l'emploi des données de la science en philosophie*, Hachette, Paris. Reprinted as Volume V (Cl. Salomon-Bayet, editor) of the Oeuvres Complètes, 1978, Paris: J. Vrin.

de Laplace, P. (1814). *Essai philosophique sur les probabilités*, 1 edn, Courcier, Paris.

de Laplace, P. (1825). *Essai philosophique sur les probabilités*, 5 edn, Courcier, Paris. The fifth edition was the definitive edition.

de Laplace, P. (1986). *Essai philosophique sur les probabilités*, Christian Bourgois, Paris. This is the modern edition edited by Bernard Bru.

Hoffman, F. (ed.) (1998). *Symposia in Applied Mathematics*, Vol. 55, American Mathematical Society.

Leibniz, G. (1995). *L'Estime des apparences*, J. Vrin, Paris. Edited by Marc Parmentier.

Lieberson, S. (1985). *Making It Count: The Improvement of Social Science Research*, University of California Press, Berkeley.

Martin, T. (1996). *Probabilités et critique philsophique selon Cournot*, Vrin, Paris.

McKim, V. and Turner, S. (eds) (1996). *In Statistical Methods and the Search for Causal Knowledge in the Social Sciences*, University of Notre Dame Press.

Mellor, D. (1994). *Matters of Metaphysics*, Cambridge University Press, Cambridge.

Parmentier, M. (1995). Introduction, to (Leibniz, 1995, 7–43).

Ramsey, F. (1990). *Philosophical Papers*, Cambridge University Press, Cambridge. Edited by D. H. Mellor.

Sahlin, N.-E. (1990). *The Philosophy of F. P. Ramsey*, Cambridge University Press, Cambridge.

Saxer, W. (ed.) (1932). *Verhandlungen der Internationalen Mathematiker-Kongresses Zurich 1932*, Vol. I: Berichte und Allgemeine Vorträge, Orell Füssli Verlag, Zürich.

Shafer, G. (1996a). *The Art of Causal Conjecture*, MIT Press, Cambridge–Massachusetts.

Shafer, G. (1996b). The significance of Jacob Bernoulli's Ars Conjectandi for the philosophy of probability today, *Journal of Econometrics* **75**: 15–32.

Shafer, G. (1998). Mathematical foundations for probability and causality, in (Hoffman, 1998, 207–270).

Smith, C. and Wise, M. N. (1989). *Energy and Empire: A Biographical Study of Lord Kelvin*, Cambridge University Press, Cambridge.

Woods, M. (1997). *Conditionals*, Clarendon Press, Oxford.

T. SEIDENFELD

REMARKS ON THE THEORY OF CONDITIONAL
PROBABILITY: SOME ISSUES OF FINITE VERSUS
COUNTABLE ADDITIVITY[1]

1. INTRODUCTION

Why does it matter whether probability is countably additive or merely finitely additive?

Recall the received view of mathematical probability. Let $\mathcal{B}$ be a σ-field of sets of subsets of Ω, with points ω. Ordinary elements of $\mathcal{B}$ are denoted by upper-case letters, 'E', 'F', etc. Kolmogorov's 1933 (1950) axiomatization of probability requires that $\forall (A, B) \in \mathcal{B}$:

AXIOM 1. $0 < P(A) < 1$.
AXIOM 2. $P(\Omega) = 1$.
AXIOM 3. If $A \cap B = \emptyset$ then $P(A) + P(B) = P(A \cup B)$.

Last, σ-additivity is taken by Kolmogorov as an "expedient" (1950, 15):

AXIOM 4. If $(A_i \cap A_j) = \emptyset$ whenever $i \neq j$ $(i, j = 1, 2, \ldots)$ then

$$P(\cup_i A_i) = \sum_i P(A_i).$$

Here are six reasons for considering the theory of finitely additive probability: the theory that results from Axioms 1-3, without requiring Axiom 4.

REASON 1. Measurability precludes non-trivial, σ-additive unconditional probabilities defined over the powerset of Ω when that is an uncountable set. (See Ulam's theorem, e.g., (Jech, 1978, 297)) That is, with the received theory, the domain of probability is restricted to a proper sub-σ-field of an uncountable Ω. However, an application of the Hahn-Banach theorem establishes that probability can always be extended to the power set if it is finitely additive (see Ash (1972)). The interesting work of Dubins and Savage (1976) relies on just this flexibility of finitely additive probability to avoid problems of measurability. However, this maneuver opens the door to an interesting debate over non-constructive methods in probability since, by an important result due to Solovay, without the axiom of choice, using the axiom of dependent choice instead and a large cardinal assumption, all subsets of the continuum can be made Lebesgue measurable (see (Jech, 1978, 537)).

REASON 2. Limits of relative frequencies need not satisfy the axiom of countable additivity. For example, let the sample space be a countable set, $\Omega = \{\omega_1, \omega_2, \ldots\}$. Define the probability of an event E, $P(E)$, as the limit of relative frequency of E in a denumerable sequence S of repeated trials, with outcomes from Ω, $S = <o_1, o_2, \ldots>$, $o_i \in \Omega$, $i = 1, 2, \ldots$. Then, for each $\omega \in \Omega$, it can be that $o_i = \omega$ for only finitely many values of i, i.e., only finitely often does ω occur in the sequence S. Then, for each $\omega \in \Omega$, $P(\omega) = 0$, in violation of Axiom 4. See Kadane and O'Hagen (1995) for an interesting discussion of how such a finite, but not countably additive probability can be used to model selecting a natural number "at random."

REMARK 1. The class of events for which the limit of frequency is defined need not form even a field (see (Billingsley, 1986, problem 2.15.)). However, the same argument cited above, involving the Hanh-Banach theorem, shows that limits of relative frequencies can be extended to form a finitely additive probability on the powerset of Ω.

REASON 3. Some important decision theories require no more than finite additivity. e.g., deFinetti (1974), and Savage's (1954) theories. However, these theories require a controversial assumption about the state-independent utility for *consequences* (see Seidenfeld and Schervish (1983) and Schervish et al. (1990)).

REASON 4. Two-person, zero-sum games with bounded payoffs that do not have (minimax) solutions using σ-additive probability do, when finitely additive mixed strategies are permitted. For example, Wald's (1950) zero-sum game of picking the bigger integer, with $+1$ to the winner, -1 to the loser, and 0 in case of a tie, has no value within the class of σ-additive strategies. However (Schervish and Seidenfeld, 1996), the game is "fair," i.e. each player's minimax strategy has value 0, when purely finitely additive strategies are used. Wald's treatment of statistical games with infinite parameter spaces often leads to maximin strategies for Nature that, as limits of countably additive strategies, are purely finitely additive. These priors are the "least favorable" possible from the statistician's point of view (see, e.g, (Berger, 1985, 350)).

REASON 5. Textbook, classical statistical methods have (extended) Bayesian models that rely on purely finitely additive prior probabilities. For example, Sir Harold Jeffreys' (1971) Bayes-models of inferences for locations/scale parameters rely on "improper" priors. For example, his theory adopts $d\mu$ the uniform density for Lebesgue measure as the "improper" prior for inference about the mean, μ, based on Normal $\mathbf{N}(\mu, 1)$ data. This prior is "improper" as it does not integrate to 1 (or any quantity) though, of course, Lebesgue meaure is σ-finite. Nonetheless, when it is used in formal Bayes calculations with the likelihood from the Normal $\mathbf{N}(\mu, 1)$ model, the resulting posterior distributions are proper and agree with the nominal Confidence Intervals for the same inference problem. However, this "improper" prior assigns equal (finite) weight to each bounded interval of length k for μ. Stated in terms of probabilities, this "improper" prior gives equal probability to each unit interval for μ. As the parameter space is a countable union of disjoint unit intervals, the "improper" prior density, $d\mu$, corresponds to a (class of) finitely but not countably additive prior probability distribution(s) on μ.

REASON 6. When $P(A) > 0$, $A \in \mathcal{B}$, then conditional probability given A is well defined by

$$P(\cdot|A) = P(\cdot \cap A)/P(A).$$

The focus in the balance of this paper is on some differences between the two theories, with and without the axiom of countable additive, for the development of conditional probability with respect to null events, when $P(A) = 0$.

2. CONDITIONAL PROBABILITY $P(\cdot|A)$ WHEN $P(A) = 0$

Let $<\Omega, \mathcal{B}, P>$ be a measure space. Denote by ω points in Ω. When $P(A) > 0$, $A \in \mathcal{B}$, then conditional probability over $\mathcal{B}$ given A is well defined by the familiar rule, $P(\cdot|A) = P(\cdot \cap A)/P(A)$. Of course, this approach does not provide guidance for conditional probability given P-null (measure 0) events. For that, the received view comes from Kolmogorov's seminal 1933 work. In the usual terminology, with $\mathcal{A}$ a sub-σ-field of $\mathcal{B}$,

$P(\cdot|\mathcal{A})$ is a *regular conditional distribution* [rcd] on $\mathcal{B}$, given $\mathcal{A}$ provided that:
 1. For each $\omega \in \Omega$, $P(\cdot|\mathcal{A})(\omega)$ is a probability on $\mathcal{B}$.
 2. For each $B \in \mathcal{B}$, $P(B|\mathcal{A})(\cdot)$ is an $\mathcal{A}$-measurable function.
 3. For each $A \in \mathcal{A}$, $P(A \cap B) = \int_A P(B|\mathcal{A})(\omega)dP(\omega)$.

That is, $P(B|\mathcal{A})$ is a version of the Radon-Nikodym derivative of $P(\cdot \cap B)$ with respect to $P(\cdot)$.

2.1. Two limitations in this approach are well documented

- An rcd may not exist.

The canonical example of a measure space that admits no rcd's is obtained by extending the σ-field of Borel sets on $[0, 1]$ under Lebesgue measure with the addition of one non-measurable set. Denote the original measure space by $<[0,1], \mathcal{A}, \lambda>$. A familiar maneuver allows an extension to an enlarged measure space, denoted $<\Omega, \mathcal{B}, P>$. However, there is no rcd $P(\cdot|\mathcal{A})(\omega)$ on $\mathcal{B}$ given $\mathcal{A}$. (See (Halmos, 1950, 211), (Billingsley, 1986, Exercise 33.13), (Breiman, 1968, 81), (Doob, 1953, 624), or (Loeve, 1955, 370) for variations on this common theme.) Though, for each $B \in \mathcal{B}$, the extended measure space has Radon-Nikodym derivatives $P(B|\cdot)$ satisfying condition 3, above, these resist assembly of pointwise probabilities into a full probability distribution on $\mathcal{B}$, as required by condition 1.

In recent work (Seidenfeld et al., 2000) we show that, quite generally, a measure space admitting rcd's can be extended to another measure space admitting rcd's if and only if the latter lies within the measure completion of the former. This finding hints at other links between (a) the aforementioned problem of nonmeasurable sets for an unconditional probability and (b) the theory of conditional probability distributions when Axiom 4 is imposed. We discuss that in our paper (2000).

In rejoinder to the existence problem, however, a sufficient condition for rcd's to exist on $\mathcal{B}$ (given any sub σ-field $\mathcal{A}$) is that $\mathcal{B}$ be isomorphic under a 1-1 measurable mapping to the σ-field of a random variable (see, (Billingsley, 1986, T.33.3) or (Breiman, 1968, T. 4.30)).

- As Kolmogorov notes $P(\cdot|\mathcal{A})$ is *not* probability given an *event*.

He illustrates this with the so-called "Borel" paradox. Put simply, it is that $P(\cdot|\mathcal{A})$ is *not* probability given an *event* but, rather, probability given a *σ-field*. Specifically, with $\mathcal{B}_Z$ a σ-field generated by the random variable Z, let sub-σ-fields $\mathcal{A}_X$ and $\mathcal{A}_Y$ be generated by the random variables X and Y, respectively. Suppose that $X = x$ is the same event (in $\mathcal{B}_Z$) as $Y = y$. Nonetheless, when $X(\omega) = x$ the two rcd's, $P(\cdot|\mathcal{A}_X)(\omega)$ and $P(\cdot|\mathcal{A}_Y)(\omega)$, may be different, with sup norm arbitrarily close to 1.

In rebuttal, Kolmogorov points out that between any two conditioning sub-σ-fields, this "paradox" can occur only on a P-null set of points. That is, it is a measure-0 failure, at worst.

2.2. Improper rcd's.

The focus for the balance of this section comes from important work by Blackwell and Ryll-Nardzewski (1963) and Blackwell and Dubins (1975)) (theorems reported here that are not given explicit citations can be found in Seidenfeld et al. (2000)).

DEFINITION 1.
- An rcd $P(\cdot|\mathcal{A})(\omega)$ on $\mathcal{B}$ given $\mathcal{A}$, is *proper at ω* if $P(\cdot|\mathcal{A})(\omega) = 1$ whenever $\omega \in A \in \mathcal{A}$.
- $P(\cdot|\mathcal{A})(\omega)$ is *improper at ω*, otherwise.
- $P(\cdot|\mathcal{A})$ is proper, if $P(\cdot|\mathcal{A})(\omega)$ is proper at every $\omega \in \Omega$.

Two other useful concepts for the presentation here are these:
- An *$\mathcal{A}$-atom* is the intersection of all elements of $\mathcal{A}$ that contain given point ω of Ω.

- A probability distribution is *extreme* if its range is the two point set $\{0,1\}$.

THEOREM 1. Blackwell and Dubins (1975). When $\mathcal{B}$ is a countably generated σ-field, *no* rcd on $\mathcal{B}$ given $\mathcal{A}$ is proper if there exists *some* extreme probability on $\mathcal{A}$ supported by no $\mathcal{A}$-atom belonging to $\mathcal{A}$.

In other words, provided there exists one extreme probability on $\mathcal{A}$ which is not supported by its $\mathcal{A}$-atoms, then the sub-σ-field $\mathcal{A}$ is anomalous for all rcd's on $\mathcal{B}$ given $\mathcal{A}$ in that they are improper, each and every one!

Now, for our central result about the extent of impropriety in certain rcd's. Assume that $\mathcal{A}$ is an atomic sub-σ-field of $\mathcal{B}$, with $\mathcal{A}$-atoms a. Denote by $a(\omega)$ that $\mathcal{A}$-atom containing the point ω.

THEOREM 2. Let P be an extreme probability on $\mathcal{A}$ that is not supported by any of its $\mathcal{A}$-atoms. If an rcd $P(\cdot|\mathcal{A})(\omega)$ on $\mathcal{B}$ given $\mathcal{A}$ exists, there is one where $P\{\omega : P(a(\omega)|\mathcal{A})(\omega)) = 0\} = 1$. And, if $\mathcal{B}$ is countably generated, then this rcd is unique.

Theorem 2 asserts that when $\mathcal{B}$ is countably generated, almost surely with respect to P, the rcd's on $\mathcal{B}$ given $\mathcal{A}$ are maximally improper, in two senses simultaneously:

1. The set of points where propriety fails has measure 1 under P.
2. For $P(a(\omega)|\mathcal{A})(\omega) = 0$ when propriety requires that
$$P(a(\omega)|\mathcal{A})(\omega) = 1.$$

Here are several examples of Theorem 2.

EXAMPLE 1. (See (Billingsley, 1986, E33.11)). Let $\Omega = [0,1]$, let $\mathcal{B}$ = the Borel subsets of Ω, and let P be Lebesgue measure. Let $\mathcal{A}$ be the sub-σ-field of all countable and co-countable sets in $[0,1]$. Clearly, $P(A) = 0$ or $P(A) = 1$, for each $A \in \mathcal{A}$. Equally obvious, $P(A) = 0$ for each countable set A. Note that the $\mathcal{A}$-atoms, which in fact belong to $\mathcal{A}$, are just the points of Ω, $\{x : 0 \le x \le 1\}$. Hence, according to Theorem 2, the rcd on $\mathcal{B}$ given $\mathcal{A}$ satisfies:
$$P\{x : P(x|\mathcal{A})(x) = 0, \text{ for } 0 \le x \le 1\} = 1.$$

EXAMPLE 2. (See (Blackwell and Dubins, 1975, 742)). Let $\Omega = \{0,1\}^{\aleph_0}$ – the sample space of infinite binary sequences. Let $\mathcal{B}$ = the Borel subsets of Ω. And let P be the product measure, corresponding to independent "fair" coin flips. Let $\mathcal{A}$ be the tail σ-field for this process. Then, by the Kolmogorov 0-1 law, for $A \in \mathcal{A}$, $P(A) = 0$ or $P(A) = 1$. $\mathcal{A}$ is atomic. The $\mathcal{A}$-atoms, a, are countable sets of points, $\omega \in \Omega$, where ω', $\omega \in a$ if they differ in at most finitely many places. Since each $\mathcal{A}$-atom is a countable set, $P(a) = 0$; hence, P is not supported by any of its $\mathcal{A}$-atoms. Thus, $P\{\omega : P(a(\omega)|\mathcal{A})(\omega) = 0\} = 1$.

Example 2 has a generalization to i.i.d. binomial "weighted" coin flipping, $P_\theta(\{1 \times \{0,1\} \times \ldots \}) = \theta, 0 < \theta < 1$, included in the next result, Theorem 3, dealing with symmetric measures, as covered by deFinetti's representation theorem for exchangeable processes.

EXAMPLE 3. (Continuing example 2). Let $\Omega = \{0,1\}^{\aleph_0}$; let $\mathcal{B}$ = the Borel subsets of Ω; and let P be a symmetric probability, in the sense of Hewitt and Savage (1955) defined as follows. Let T be an arbitrary finite permutation of the positive integers, i.e., a permutation of the coordinates of Ω that leaves all but finitely many places fixed. For $B \in \mathcal{B}$, given T, define the set $T^{-1}B = \{\omega : T(w) \in B\}$.

> P is called a *symmetric probability* if $P(T^{-1}B) = P(B)$, for each $B \in \mathcal{B}$ and T. If $B = T^{-1}B$ for all (finite) permutations T, B is called a *symmetric event*.

Let $\mathcal{A}$ be the sub-σ-field of $\mathcal{B}$ generated by the class **T** of all finite permutations of the coordinates of Ω, i.e., $\mathcal{A}$ is the σ-field of the symmetric events. $\mathcal{A}$ is atomic, with $\mathcal{A}$-atoms comprised by a countable set of sequences, each pair of sequences in the same atom differing by some finite permutation of its coordinates. In all but two cases the $\mathcal{A}$-atoms are countably infinite sets; the two exceptions are the two constant sequences.

THEOREM 3. Each rcd $P(\cdot|\mathcal{A})$ on $\mathcal{B}$ given $\mathcal{A}$, for a symmetric probability P, satisfies $P\{\omega : P(a(\omega)|\mathcal{A})(\omega) = 0\} = 1$, provided that $P(<0,0,0,\ldots>) = P(<1,1,1,\ldots>) = 0$.

REMARK 2. The proof of Theorem 3 relies on three considerations:

1. *The Hewitt-Savage 0-1 law for i.i.d. probability P_θ over the field of symmetric events – to establish that such an i.i.d probability is extreme and is not supported by $\mathcal{A}$-atoms whenever $0 < \theta < 1$. Hence, by Theorem 2, the rcd's for $P_\theta(\cdot|\mathcal{A})(\omega)$ on $\mathcal{B}$ given $\mathcal{A}$ are maximally improper.*
2. *DeFinetti's representation of symmetric probabilities as a mixture of i.i.d probabilities P_θ;*
3. *A representation of the rcd for a symmetric probability $P(\cdot|\mathcal{A})(\omega)$ on $\mathcal{B}$ given $\mathcal{A}$ as a mixture, depending upon ω, of the rcd's $P_\theta(\cdot|\mathcal{A})(\omega)$.*

To the extent that the rcd's for these examples are maximally improper, they cannot serve as coherent conditional probability distributions given the conditioning sub-σ-field—their conditional probability distributions are not supported by their conditioning events! Thus, we see that the received σ-additive theory of regular conditional distributions suffers some faults that prevent it from serving as a fully general account of degrees of belief for hypothetical reasoning when conditioning on certain (non-countably generated) sub-σ-fields.

3. SOME FINITELY ADDITIVE CONDITIONAL PROBABILITY

In dealing with finitely additive conditional probabilities $P(\cdot|\cdot)$, especially for conditional probability given events of unconditional probability 0, we adopt the following restriction:

Principle of Conditional Coherence (see Dubins (1975)):

For all pairs of events, A and B such that $A \cap B \neq \emptyset$,
1. $Q(\cdot) = P(\cdot|A)$ is a finitely additive probability.
2. $Q(A) = 1$, and $Q(\cdot|B) = P(\cdot|A \cap B)$.

When $P(A \cap B) > 0$, this principle applies, trivially. It formalizes the idea deFinetti (1974) had of conditional probability *given an event*, rather than *given a field*. That is, $P(\cdot|A)$ does not depend upon how A^c is partitioned. Hence, there is no "Borel" paradox to solve. Moreover, by clause (2), each coherent finitely additive conditional probability is proper. Last, Dubins (1975) shows that each unconditional finitely additive probability P carries a full set of coherent conditional probabilities.

What, then, is special about finitely additive conditional probability? Consider the following:

EXAMPLE 4. (attributed to P. Lévy, by deFinetti): Let P be a finitely additive probability on the denumerable set of all pairs $< s, t >$, for s and t positive integers, with the following two restrictions:

1. $P(< s, t >) = 0$, so that P is 0 on finite sets of pairs—P is purely finitely additive;
2. $P(< s, t > |B) = 0$ if B is an infinite set, so that the conditional $P(\cdot|B)$ is also purely finitely additive when B is an infinite set.

Define the events:
$$E = \{< s, t >: s > t\},$$
$$S_m = \{< s, t >: s = m\} \ (m = 1, \ldots)$$
and
$$T_n = \{< s, t >: t = n\} \ (n = 1, \ldots).$$
Then, $P(E|S_m) = 0$ for $m = 1, \ldots,$ yet $P(E|T_n) = 1$ for $n = 1, \ldots.$

Evidently the following principle is invalid for the finitely additive probability P, though it is satisfied by all σ-additive probabilities. Let $\Pi = \{h_n : n = 1, \ldots\}$ be an exhaustive partition and assume that P is defined on the field $\mathcal{B}$.

- Principle of ($\aleph_0$-)Conglomerability for Events (de Finetti):

$\forall (A \in \mathcal{B})$: If $c_1 = P(A|h_n) = c_2 \ (n = 1, \ldots)$, then $c_1 = P(A) = c_2$.

Note that in Example 4, conglomerability fails in at least one of the two orthogonal partitions, $\Pi_1 = \{S_m : m = 1, \ldots\}$ and $\Pi_2 = \{T_n : n = 1, \ldots\}$. Where conglomerability fails, there the finitely additive probability P fails also to satisfy condition 3 for regular conditional probabilities. That is, then $P(A) \neq$

$\int_\Omega P(A|h_n)dP(h)$, and P is not *disintegrable* in the partition Π. In an earlier study (Schervish et al., 1984) we showed that:

THEOREM 4. Each merely finitely additive probability fails conglomerability in some denumerable partition.

REMARK 3. Unless the probability assumes only finitely many values, there exists such a partition where conglomerability fails, whose elements each have positive prior probability, unlike in the Lévy example.

Thus, as highlighted here, the trade-off between the two theories of conditional probability can be summarized this way. With a countably additive probability, when conditioning on sub-σ-fields that are not countably generated (e.g., the field of symmetric events), the risk is that the rcd's may be maximally improper. By contrast, with a finitely additive probability, though conditional probability always is proper – the conditional probability is coherent – the risk is that the conditional probability will fail to be conglomerable (will fail to be disintegrable) in the partition of interest.

Recall that each countably additive space $<\Omega, \mathcal{B}, P>$ can be extended to a coherent finitely additive probability on the powerset of Ω. In case the rcd's $P(\cdot|\mathcal{A})$ on $\mathcal{B}$ given $\mathcal{A}$ are improper, the relevant question is whether there exists a finitely additive extension Q of P that is conglomerable (i.e., disintegrable) in the partition of the $\mathcal{A}$-atoms. If so, then one may blend the familiar properties of P, with the coherence of Q. That is, when $P(B) > 0$, then one may calculate using P's (improper) rcd's in the usual way, $P(B) = \int_\Omega P(B|\mathcal{A})(\omega)dP(\omega)$, and know that this agrees with $Q(B)$. However, for small events when $P(B) = 0$, then the propriety of Q's conditional probabilities ensures that, e.g., $Q(B|a(\omega))$ is coherent. In such a case, say that Q is a useful finitely additive extension of P.

Dubins (1977) shows that useful finitely additive extensions of Lebesgue measure exist for the case of the tail-field, Example 2. It is an open question, I believe, under what conditions such useful finitely additive probabilities exist generally.

Department of Philosophy
Carnegie Mellon University
USA

NOTES

[1] The paper is based on collaborated research with J.B. Kadane and M.J. Schervish, Department of Statistics, Carnegie Mellon University. That research was supported by NSF Grant DMS 9801401.

REFERENCES

Ash, R. B. (1972). *Real Analysis and Probability*, Academic Press, New York.

Berger, J. (1985). *Statistical Decision Theory and Bayesian Analysis*, 2nd edn, Springer, New York.

Billingsley, P. (1986). *Probability and Measure*, 2nd edn, John Wiley, New York.

Blackwell, D. and Dubins, L. E. (1975). On existence and non-existence of proper, regular, conditional distributions, *Annals of Probability* **3**: 741–752.

Blackwell, D. and Ryll-Nardzewski, C. (1963). Non-existence of everywhere proper conditional distributions, *Annals of Mathematical Statistics* **34**: 223–225.

Breiman, L. (1968). *Probability*, Addison-Wesley, Reading MA.

deFinetti, B. (1974). *The Theory of Probability*, John Wiley, New York.

Doob, J. (1953). *Stochastic Processes*, John Wiley, New York.

Dubins, L. (1975). Finitely additive conditional probabilities, conglomerability, and disintegrations, *Ann. Prob.* **3**: 89–99.

Dubins, L. (1977). Measurable, tail distintegrations of the Haar integral are purely finitely additive, *Proc. Amer. Math. Soc.* **62**: 34–36.

Dubins, L. and Savage, L. (1976). *Inequalities for Stochastic Processes: How to gamble if you must*, Dover, New York.

Halmos, P. (1950). *Measure Theory*, van Nostrand, New York.

Hewitt, E. and Savage, L. (1955). Symmetric measures on Cartesian products, *Amer. Math. Soc. Trans.* **80**: 470–501.

Jech, T. (1978). *Set Theory*, Academic Press.

Jeffreys, H. (1971). *The Theory of Probability*, 3rd edn, Oxford Univ. Press, Oxford.

Kadane, J. and O'Hagen, A. (1995). Using finitely additive probability: Uniform distributions on the natural numbers, *J. Amer. Stat. Assoc.* **90**: 626–631.

Kolmogorov, A. (1950). *Foundations of the Theory of Probability*, Chelsea, New York.

Loeve, M. (1955). *Probability Theory*, van Nostrand, New York.

Savage, L. (1954). *The Foundations of Statistics*, John Wiley, New York.

Schervish, M. and Seidenfeld, T. (1996). A fair minimax theorem for two-person (zero sum) games involving finitely additive strategies, in *Bayesian Analysis in Statistics and Econometrics*. Eds. D. Berry, C. Chaloner, and J. Geweke. New York: John Wiley.

Schervish, M., Seidenfeld, T. and Kadane, J. (1984). The extent of non-conglomerability of finitely additive probability, *Z. War.* **66**: 205–226.

Schervish, M., Seidenfeld, T. and Kadane, J. (1990). State-dependent utilities, *J. Amer. Stat. Assoc.* **85**: 840–847.

Seidenfeld, T. and Schervish, M. (1983). A conflict between finite additivity and avoiding dutch book, *Phil. Sci.* **50**: 398–412.

Seidenfeld, T., Schervish, M. and Kadane, J. (2000). Improper regular conditional distributions, Technical Report, Dept. of Statistics, Carnegie Mellon University.

Wald, A. (1950). *Statistical Decision Functions*, John Wiley, New York.

INDEX

additivity
  σ-, 167
algorithmic information theory, 39
algorithmic theory of randomness, 51
analysis
  time-series, 112
applications, 73
Ash, R.B., 167
Asmussen, S., 28
axiom of randomness, 81
axiomatization, 72

Banach, S., 17
Barnard, G.A., 25
Bayes, T., 24
Bayesian
  inference, 126, 138
  network, 143
Bernard, C., 164
Bernoulli, J., 152
Bernstein polynomial, 133
Bernstein, G., 17
Bernstein, S.N., 162
Bertrand, J., 71
betting behaviour, 77

Bickel, B.J., 23
Billingsley, P., 23, 168
Birnstein, A., 25
Black-Scholes formula, 37
Blackwell, D., 171
Blind Chance, 153
Boltzmann, L., 31
bootstrap, 35
Borel, E., 71
Bradley, F.H., 107, 108
Breiman, L., 170
Brown, R., 19
Brownian motion, 19
Bru, B., 163

calculus
  Itô, 20
  Palm, 28
  stochastic, 28
Cantelli, F.P., 17, 23
Carnap's continuum of inductive methods, 141
Cartwright, N., 158
causality, 161
Chrystal, G., 117
Church, A., 55, 81

Cohen, J.W., 28
collectives, 39, 55
Comte, A., 164
condition A, 53
condition B, 53
Cournot, A.A., 154
Cox, D.R., 30
Cramér, H., 18

de Finetti, B., 17
de Laplace, P. S., 154, 163
decision theory, 26, 77
deficiency of randomness, 58
deFinetti, B., 168
  representation theorem, 173
Dellacherie, C., 20
denumerable probabilities, 75
derivative
  Radon-Nikodym, 170
determinism, 90, 155, 161
  on different scales, 91
disintegrable, 176
distribution
  Dirichlet, 137
  natural, 132, 135, 139
  regular conditional, 170
  uniform, 128, 129, 137, 138
Doob, J.L., 18, 170
Dubins, L.E., 167, 171
Dudley, R.M., 23

Edgeworth, F.Y., 21, 108
Einstein, A., 31
Ellis, L., 155
Erlang, K.A., 28
event
  symmetric, 173
exchangeable distributions, 62
exchangeable processes, 173
expert system, 125, 138

Feller, W., 17
Feynman, R.P., 33
filtration, 20

finance, 36
Fisher, R.A., 21
free will, 161
frequencies
  limits of relative, 168
frequency interpretation of probability, 87
frequency interpretation of probabiliy, 97

Galton, F., 21
game
  two-person, zero-sum, 26, 168
Gibbs, J.W., 31
Glivenko, V., 23
God, 155, 162

Hacking, I., 16, 114
Haldane, J.B.S., 21, 29
Halmos, P.R., 170
Hardy, G.H., 29
Hewitt, E., 173
Hewitt-Savage 0-1 law, 174
Hilbert, D., 72

indeterminism, 90, 161
indifference, 128, 138
individualistic mentality, 75
inference
  inductive, 87
Integral
  Itô, 20
integral
  stochastic, 20

Jeffrey, H., 169
Jevons, W.S., 113, 114
Johnson's Sufficientness Postulate, 141
Johnson, W.E., 117

Kadane, J.B., 168, 177
Keynes, J.M., 71, 109
Khinchin, S., 17
Kolmogorov complexity, 56

INDEX

Kolmogorov, A.N., 17, 39, 72
  Grundbegriffe, 17
  0-1 law, 173
  axiomatization of probability, 167
  complexity conception of probability, 51
  frequency interpretation, 52

Lévy, P., 175
Lauritzen, S., 35
law
  of large numbers, 23
  of nature, 90
  physical, 91
law of the iterated logarithm, 55
laws of chance, 71
Leibniz, G.W.F., 163
Lieberson, S., 160
Lindley, D.V., 26
Loeve, M., 170

Malthus, R., 113
Marcinkiewicz, R., 17
marginalization, 125, 138
Markov Chain, 35
Markov Chain Monte Carlo (MCMC), 35
Markov, A.A., 17
Markowitz, H., 36
Martin, T., 163
Martin-Löf, P., 61
martingales, 61
mathematical
  certitude, 91
mathematical genetics, 29
maximin, 169
Maxwell, J.C., 31, 164
measure theory
  the rise of, 16
  the triumph of, 17
Mellor, D.H., 156
Meyer, P.-A., 20
Mill, J.S., 73, 99

minimax, 168
model
  complexity, 58
  statistical, 58
modeling, 125
Monte Carlo, 34
Morgenstein, O., 26

natural
  distribution, 126
  probability function, 126
  random process, 127
Nature, 147
  active, 147
  as a limit, 152
  as a witness, 159
  as an actor, 147
  as an idealization, 151
  passive, 147
Nature's
  event tree, 149, 150
  expectations, 147
  laws, 147
  possibilities, 147
  predictions, 147
  tree, 158
Newton-Leibniz calculus, 20
non-measurable sets, 76

O'Hagen, A., 168
objective probabilities, 78, 155
  as subjective, 156
objects
  stochastic, 61
options
  vanilla, 38

Paley, R.E.A.C, 42
Palm, C., 28
paradox
  Borel, 171, 175
Parmentier, M., 164
Pascal, B., 152
Pearson, E., 135

181

Pearson, K., 21
Peirce, C.S., 105, 106, 154
Poincaré, H., 71
Poisson, S.-D., 111
Porter, T., 114
posits, 87
practical continuum, 75
practical value, 73
pre-selection, 131, 132
prediction, 88
principle
  ($\aleph_0$-)Conglomerability for Events, 175
  conditional coherence, 174
  Erdös-Kac-Donsker invariance, 23
  likelihood, 25
prior
  "improper", 169
  "least favorable", 169
  Dirichlet, 138
  problem, 126
  uniform, 135
probabilities
  imprecise, 126
probability
  applied, 27
  conditional, 167
  finitely additive, 167
  space filtered, 20
  symmetric, 173
process
  0-1 random, 129
  random, 153
  stochastic, 18, 156

quantum mechanics, 162
quantum theory, 32
Quetelet, A., 111–113
queuing theory, 28

Ramsey, F.P., 26, 154, 164
random sequences, 55
rates, 33

realities, 73
reasoning
  defeasible, 128
reference class problem, 97
regularities
  causal, 148, 156, 160
  dynamic, 148
Reichenbach, H., 71
relations
  causal, 148
reliability theory, 30
Rogers, L.C.G., 21
Ryll-Nardzewski, C., 171

Sahlin, N.-E., 164
Savage, L.J., 22, 167, 173
Schervish, M.J., 168, 177
science of prices and exchange, 85
Seidenfeld, T., 22
selection rules, 55
sequence
  Bernoulli, 57
  Gaussian, 59
  Markov, 59
  Poisson, 60
Shafer, G., 41
Sheynin, O.B., 16
Shorack, G.R., 23
Solovay, R., 168
statistical mechanics, 31
statistics
  Bayesian, 24, 153
  classical, 153
  non-parametric, 22
  parametric, 22
  semi-parametric, 23
Steinhaus, T., 17
Stigler, S., 97
stochastic mechanism, 153
subjective probabilities, 78, 152
survival analysis, 30
symmetry, 129, 141

tail σ-field, 173

Theorem
    Glivenko-Cantelli, 23
theorem
    Bayes', 24
    de Finetti's, 63
    Hahn-Banach, 167
    Kolmogorov-Smirnov, 23
    Ulam's, 167
Thomson, W.T., 164
tree
    event, 149
    probability, 149, 158
turbulence, 33
Turing machine
    universal, 56

unity of sciences, 71

van der Waart, A.V., 23
Venn, J., 97, 155
volatility, 37
von Mises, R., 39, 71, 115
von Neumann, J., 26
Vovk, V., 41

Wald, A., 168
weak renaming, 141
Wellner, J.A., 23
Wiener, N., 17, 20
Williams, D., 21
Wittgenstein, L., 41
Wood, M., 156
Wright, S., 21

Zygmund, F., 17

SYNTHESE LIBRARY

1. J. M. Bochénski, *A Precis of Mathematical Logic*. Translated from French and German by O. Bird. 1959 ISBN 90-277-0073-7
2. P. Guiraud, *Problèmes et méthodes de la statistique linguistique*. 1959 ISBN 90-277-0025-7
3. H. Freudenthal (ed.), *The Concept and the Role of the Model in Mathematics and Natural and Social Sciences*. 1961 ISBN 90-277-0017-6
4. E. W. Beth, *Formal Methods*. An Introduction to Symbolic Logic and to the Study of Effective Operations in Arithmetic and Logic. 1962 ISBN 90-277-0069-9
5. B. H. Kazemier and D. Vuysje (eds.), *Logic and Language*. Studies dedicated to Professor Rudolf Carnap on the Occasion of His 70th Birthday. 1962 ISBN 90-277-0019-2
6. M. W. Wartofsky (ed.), *Proceedings of the Boston Colloquium for the Philosophy of Science, 1961–1962*. [Boston Studies in the Philosophy of Science, Vol. I] 1963 ISBN 90-277-0021-4
7. A. A. Zinov'ev, *Philosophical Problems of Many-valued Logic*. A revised edition, edited and translated (from Russian) by G. Küng and D.D. Comey. 1963 ISBN 90-277-0091-5
8. G. Gurvitch, *The Spectrum of Social Time*. Translated from French and edited by M. Korenbaum and P. Bosserman. 1964 ISBN 90-277-0006-0
9. P. Lorenzen, *Formal Logic*. Translated from German by F.J. Crosson. 1965 ISBN 90-277-0080-X
10. R. S. Cohen and M. W. Wartofsky (eds.), *Proceedings of the Boston Colloquium for the Philosophy of Science, 1962–1964*. In Honor of Philipp Frank. [Boston Studies in the Philosophy of Science, Vol. II] 1965 ISBN 90-277-9004-0
11. E. W. Beth, *Mathematical Thought*. An Introduction to the Philosophy of Mathematics. 1965 ISBN 90-277-0070-2
12. E. W. Beth and J. Piaget, *Mathematical Epistemology and Psychology*. Translated from French by W. Mays. 1966 ISBN 90-277-0071-0
13. G. Küng, *Ontology and the Logistic Analysis of Language*. An Enquiry into the Contemporary Views on Universals. Revised ed., translated from German. 1967 ISBN 90-277-0028-1
14. R. S. Cohen and M. W. Wartofsky (eds.), *Proceedings of the Boston Colloquium for the Philosophy of Sciences, 1964–1966*. In Memory of Norwood Russell Hanson. [Boston Studies in the Philosophy of Science, Vol. III] 1967 ISBN 90-277-0013-3
15. C. D. Broad, *Induction, Probability, and Causation*. Selected Papers. 1968 ISBN 90-277-0012-5
16. G. Patzig, *Aristotle's Theory of the Syllogism*. A Logical-philosophical Study of *Book A* of the *Prior Analytics*. Translated from German by J. Barnes. 1968 ISBN 90-277-0030-3
17. N. Rescher, *Topics in Philosophical Logic*. 1968 ISBN 90-277-0084-2
18. R. S. Cohen and M. W. Wartofsky (eds.), *Proceedings of the Boston Colloquium for the Philosophy of Science, 1966–1968, Part I*. [Boston Studies in the Philosophy of Science, Vol. IV] 1969 ISBN 90-277-0014-1
19. R. S. Cohen and M. W. Wartofsky (eds.), *Proceedings of the Boston Colloquium for the Philosophy of Science, 1966–1968, Part II*. [Boston Studies in the Philosophy of Science, Vol. V] 1969 ISBN 90-277-0015-X
20. J. W. Davis, D. J. Hockney and W. K. Wilson (eds.), *Philosophical Logic*. 1969 ISBN 90-277-0075-3
21. D. Davidson and J. Hintikka (eds.), *Words and Objections*. Essays on the Work of W. V. Quine. 1969, rev. ed. 1975 ISBN 90-277-0074-5; Pb 90-277-0602-6
22. P. Suppes, *Studies in the Methodology and Foundations of Science*. Selected Papers from 1951 to 1969. 1969 ISBN 90-277-0020-6
23. J. Hintikka, *Models for Modalities*. Selected Essays. 1969 ISBN 90-277-0078-8; Pb 90-277-0598-4

SYNTHESE LIBRARY

24. N. Rescher *et al.* (eds.), *Essays in Honor of Carl G. Hempel*. A Tribute on the Occasion of His 65th Birthday. 1969 ISBN 90-277-0085-0
25. P. V. Tavanec (ed.), *Problems of the Logic of Scientific Knowledge*. Translated from Russian. 1970 ISBN 90-277-0087-7
26. M. Swain (ed.), *Induction, Acceptance, and Rational Belief.* 1970 ISBN 90-277-0086-9
27. R. S. Cohen and R. J. Seeger (eds.), *Ernst Mach: Physicist and Philosopher*. [Boston Studies in the Philosophy of Science, Vol. VI]. 1970 ISBN 90-277-0016-8
28. J. Hintikka and P. Suppes, *Information and Inference*. 1970 ISBN 90-277-0155-5
29. K. Lambert, *Philosophical Problems in Logic*. Some Recent Developments. 1970 ISBN 90-277-0079-6
30. R. A. Eberle, *Nominalistic Systems*. 1970 ISBN 90-277-0161-X
31. P. Weingartner and G. Zecha (eds.), *Induction, Physics, and Ethics*. 1970 ISBN 90-277-0158-X
32. E. W. Beth, *Aspects of Modern Logic*. Translated from Dutch. 1970 ISBN 90-277-0173-3
33. R. Hilpinen (ed.), *Deontic Logic*. Introductory and Systematic Readings. 1971
    See also No. 152. ISBN Pb (1981 rev.) 90-277-1302-2
34. J.-L. Krivine, *Introduction to Axiomatic Set Theory*. Translated from French. 1971 ISBN 90-277-0169-5; Pb 90-277-0411-2
35. J. D. Sneed, *The Logical Structure of Mathematical Physics*. 2nd rev. ed., 1979 ISBN 90-277-1056-2; Pb 90-277-1059-7
36. C. R. Kordig, *The Justification of Scientific Change*. 1971 ISBN 90-277-0181-4; Pb 90-277-0475-9
37. M. Čapek, *Bergson and Modern Physics*. A Reinterpretation and Re-evaluation. [Boston Studies in the Philosophy of Science, Vol. VII] 1971 ISBN 90-277-0186-5
38. N. R. Hanson, *What I Do Not Believe, and Other Essays*. Ed. by S. Toulmin and H. Woolf. 1971 ISBN 90-277-0191-1
39. R. C. Buck and R. S. Cohen (eds.), *PSA 1970*. Proceedings of the Second Biennial Meeting of the Philosophy of Science Association, Boston, Fall 1970. In Memory of Rudolf Carnap. [Boston Studies in the Philosophy of Science, Vol. VIII] 1971 ISBN 90-277-0187-3; Pb 90-277-0309-4
40. D. Davidson and G. Harman (eds.), *Semantics of Natural Language*. 1972 ISBN 90-277-0304-3; Pb 90-277-0310-8
41. Y. Bar-Hillel (ed.), *Pragmatics of Natural Languages*. 1971 ISBN 90-277-0194-6; Pb 90-277-0599-2
42. S. Stenlund, *Combinators, γ Terms and Proof Theory*. 1972 ISBN 90-277-0305-1
43. M. Strauss, *Modern Physics and Its Philosophy*. Selected Paper in the Logic, History, and Philosophy of Science. 1972 ISBN 90-277-0230-6
44. M. Bunge, *Method, Model and Matter*. 1973 ISBN 90-277-0252-7
45. M. Bunge, *Philosophy of Physics*. 1973 ISBN 90-277-0253-5
46. A. A. Zinov'ev, *Foundations of the Logical Theory of Scientific Knowledge (Complex Logic)*. Revised and enlarged English edition with an appendix by G. A. Smirnov, E. A. Sidorenka, A. M. Fedina and L. A. Bobrova. [Boston Studies in the Philosophy of Science, Vol. IX] 1973 ISBN 90-277-0193-8; Pb 90-277-0324-8
47. L. Tondl, *Scientific Procedures*. A Contribution concerning the Methodological Problems of Scientific Concepts and Scientific Explanation. Translated from Czech by D. Short. Edited by R.S. Cohen and M.W. Wartofsky. [Boston Studies in the Philosophy of Science, Vol. X] 1973 ISBN 90-277-0147-4; Pb 90-277-0323-X
48. N. R. Hanson, *Constellations and Conjectures*. 1973 ISBN 90-277-0192-X

SYNTHESE LIBRARY

49. K. J. J. Hintikka, J. M. E. Moravcsik and P. Suppes (eds.), *Approaches to Natural Language.* 1973 ISBN 90-277-0220-9; Pb 90-277-0233-0
50. M. Bunge (ed.), *Exact Philosophy.* Problems, Tools and Goals. 1973 ISBN 90-277-0251-9
51. R. J. Bogdan and I. Niiniluoto (eds.), *Logic, Language and Probability.* 1973 ISBN 90-277-0312-4
52. G. Pearce and P. Maynard (eds.), *Conceptual Change.* 1973 ISBN 90-277-0287-X; Pb 90-277-0339-6
53. I. Niiniluoto and R. Tuomela, *Theoretical Concepts and Hypothetico-inductive Inference.* 1973 ISBN 90-277-0343-4
54. R. Fraïssé, *Course of Mathematical Logic* – Volume 1: *Relation and Logical Formula.* Translated from French. 1973 ISBN 90-277-0268-3; Pb 90-277-0403-1 (For *Volume 2* see under No. 69).
55. A. Grünbaum, *Philosophical Problems of Space and Time.* Edited by R.S. Cohen and M.W. Wartofsky. 2nd enlarged ed. [Boston Studies in the Philosophy of Science, Vol. XII] 1973 ISBN 90-277-0357-4; Pb 90-277-0358-2
56. P. Suppes (ed.), *Space, Time and Geometry.* 1973 ISBN 90-277-0386-8; Pb 90-277-0442-2
57. H. Kelsen, *Essays in Legal and Moral Philosophy.* Selected and introduced by O. Weinberger. Translated from German by P. Heath. 1973 ISBN 90-277-0388-4
58. R. J. Seeger and R. S. Cohen (eds.), *Philosophical Foundations of Science.* [Boston Studies in the Philosophy of Science, Vol. XI] 1974 ISBN 90-277-0390-6; Pb 90-277-0376-0
59. R. S. Cohen and M. W. Wartofsky (eds.), *Logical and Epistemological Studies in Contemporary Physics.* [Boston Studies in the Philosophy of Science, Vol. XIII] 1973 ISBN 90-277-0391-4; Pb 90-277-0377-9
60. R. S. Cohen and M. W. Wartofsky (eds.), *Methodological and Historical Essays in the Natural and Social Sciences. Proceedings of the Boston Colloquium for the Philosophy of Science, 1969–1972.* [Boston Studies in the Philosophy of Science, Vol. XIV] 1974 ISBN 90-277-0392-2; Pb 90-277-0378-7
61. R. S. Cohen, J. J. Stachel and M. W. Wartofsky (eds.), *For Dirk Struik. Scientific, Historical and Political Essays.* [Boston Studies in the Philosophy of Science, Vol. XV] 1974 ISBN 90-277-0393-0; Pb 90-277-0379-5
62. K. Ajdukiewicz, *Pragmatic Logic.* Translated from Polish by O. Wojtasiewicz. 1974 ISBN 90-277-0326-4
63. S. Stenlund (ed.), *Logical Theory and Semantic Analysis.* Essays dedicated to Stig Kanger on His 50th Birthday. 1974 ISBN 90-277-0438-4
64. K. F. Schaffner and R. S. Cohen (eds.), *PSA 1972. Proceedings of the Third Biennial Meeting of the Philosophy of Science Association.* [Boston Studies in the Philosophy of Science, Vol. XX] 1974 ISBN 90-277-0408-2; Pb 90-277-0409-0
65. H. E. Kyburg, Jr., *The Logical Foundations of Statistical Inference.* 1974 ISBN 90-277-0330-2; Pb 90-277-0430-9
66. M. Grene, *The Understanding of Nature.* Essays in the Philosophy of Biology. [Boston Studies in the Philosophy of Science, Vol. XXIII] 1974 ISBN 90-277-0462-7; Pb 90-277-0463-5
67. J. M. Broekman, *Structuralism: Moscow, Prague, Paris.* Translated from German. 1974 ISBN 90-277-0478-3
68. N. Geschwind, *Selected Papers on Language and the Brain.* [Boston Studies in the Philosophy of Science, Vol. XVI] 1974 ISBN 90-277-0262-4; Pb 90-277-0263-2
69. R. Fraïssé, *Course of Mathematical Logic* – Volume 2: *Model Theory.* Translated from French. 1974 ISBN 90-277-0269-1; Pb 90-277-0510-0 (For *Volume 1* see under No. 54).

SYNTHESE LIBRARY

70. A. Grzegorczyk, *An Outline of Mathematical Logic*. Fundamental Results and Notions explained with all Details. Translated from Polish. 1974   ISBN 90-277-0359-0; Pb 90-277-0447-3
71. F. von Kutschera, *Philosophy of Language*. 1975   ISBN 90-277-0591-7
72. J. Manninen and R. Tuomela (eds.), *Essays on Explanation and Understanding*. Studies in the Foundations of Humanities and Social Sciences. 1976   ISBN 90-277-0592-5
73. J. Hintikka (ed.), *Rudolf Carnap, Logical Empiricist*. Materials and Perspectives. 1975
    ISBN 90-277-0583-6
74. M. Čapek (ed.), *The Concepts of Space and Time*. Their Structure and Their Development. [Boston Studies in the Philosophy of Science, Vol. XXII] 1976
    ISBN 90-277-0355-8; Pb 90-277-0375-2
75. J. Hintikka and U. Remes, *The Method of Analysis*. Its Geometrical Origin and Its General Significance. [Boston Studies in the Philosophy of Science, Vol. XXV] 1974
    ISBN 90-277-0532-1; Pb 90-277-0543-7
76. J. E. Murdoch and E. D. Sylla (eds.), *The Cultural Context of Medieval Learning*. [Boston Studies in the Philosophy of Science, Vol. XXVI] 1975
    ISBN 90-277-0560-7; Pb 90-277-0587-9
77. S. Amsterdamski, *Between Experience and Metaphysics*. Philosophical Problems of the Evolution of Science. [Boston Studies in the Philosophy of Science, Vol. XXXV] 1975
    ISBN 90-277-0568-2; Pb 90-277-0580-1
78. P. Suppes (ed.), *Logic and Probability in Quantum Mechanics*. 1976
    ISBN 90-277-0570-4; Pb 90-277-1200-X
79. H. von Helmholtz: *Epistemological Writings*. The Paul Hertz / Moritz Schlick Centenary Edition of 1921 with Notes and Commentary by the Editors. Newly translated from German by M. F. Lowe. Edited, with an Introduction and Bibliography, by R. S. Cohen and Y. Elkana. [Boston Studies in the Philosophy of Science, Vol. XXXVII] 1975
    ISBN 90-277-0290-X; Pb 90-277-0582-8
80. J. Agassi, *Science in Flux*. [Boston Studies in the Philosophy of Science, Vol. XXVIII] 1975
    ISBN 90-277-0584-4; Pb 90-277-0612-2
81. S. G. Harding (ed.), *Can Theories Be Refuted?* Essays on the Duhem-Quine Thesis. 1976
    ISBN 90-277-0629-8; Pb 90-277-0630-1
82. S. Nowak, *Methodology of Sociological Research*. General Problems. 1977
    ISBN 90-277-0486-4
83. J. Piaget, J.-B. Grize, A. Szemińska and V. Bang, *Epistemology and Psychology of Functions*. Translated from French. 1977   ISBN 90-277-0804-5
84. M. Grene and E. Mendelsohn (eds.), *Topics in the Philosophy of Biology*. [Boston Studies in the Philosophy of Science, Vol. XXVII] 1976   ISBN 90-277-0595-X; Pb 90-277-0596-8
85. E. Fischbein, *The Intuitive Sources of Probabilistic Thinking in Children*. 1975
    ISBN 90-277-0626-3; Pb 90-277-1190-9
86. E. W. Adams, *The Logic of Conditionals*. An Application of Probability to Deductive Logic. 1975   ISBN 90-277-0631-X
87. M. Przełęcki and R. Wójcicki (eds.), *Twenty-Five Years of Logical Methodology in Poland*. Translated from Polish. 1976   ISBN 90-277-0601-8
88. J. Topolski, *The Methodology of History*. Translated from Polish by O. Wojtasiewicz. 1976
    ISBN 90-277-0550-X
89. A. Kasher (ed.), *Language in Focus: Foundations, Methods and Systems*. Essays dedicated to Yehoshua Bar-Hillel. [Boston Studies in the Philosophy of Science, Vol. XLIII] 1976
    ISBN 90-277-0644-1; Pb 90-277-0645-X

SYNTHESE LIBRARY

90. J. Hintikka, *The Intentions of Intentionality and Other New Models for Modalities.* 1975
ISBN 90-277-0633-6; Pb 90-277-0634-4
91. W. Stegmüller, *Collected Papers on Epistemology, Philosophy of Science and History of Philosophy.* 2 Volumes. 1977 Set ISBN 90-277-0767-7
92. D. M. Gabbay, *Investigations in Modal and Tense Logics with Applications to Problems in Philosophy and Linguistics.* 1976 ISBN 90-277-0656-5
93. R. J. Bogdan, *Local Induction.* 1976 ISBN 90-277-0649-2
94. S. Nowak, *Understanding and Prediction.* Essays in the Methodology of Social and Behavioral Theories. 1976 ISBN 90-277-0558-5; Pb 90-277-1199-2
95. P. Mittelstaedt, *Philosophical Problems of Modern Physics.* [Boston Studies in the Philosophy of Science, Vol. XVIII] 1976 ISBN 90-277-0285-3; Pb 90-277-0506-2
96. G. Holton and W. A. Blanpied (eds.), *Science and Its Public: The Changing Relationship.* [Boston Studies in the Philosophy of Science, Vol. XXXIII] 1976
ISBN 90-277-0657-3; Pb 90-277-0658-1
97. M. Brand and D. Walton (eds.), *Action Theory.* 1976 ISBN 90-277-0671-9
98. P. Gochet, *Outline of a Nominalist Theory of Propositions.* An Essay in the Theory of Meaning and in the Philosophy of Logic. 1980 ISBN 90-277-1031-7
99. R. S. Cohen, P. K. Feyerabend, and M. W. Wartofsky (eds.), *Essays in Memory of Imre Lakatos.* [Boston Studies in the Philosophy of Science, Vol. XXXIX] 1976
ISBN 90-277-0654-9; Pb 90-277-0655-7
100. R. S. Cohen and J. J. Stachel (eds.), *Selected Papers of Léon Rosenfield.* [Boston Studies in the Philosophy of Science, Vol. XXI] 1979 ISBN 90-277-0651-4; Pb 90-277-0652-2
101. R. S. Cohen, C. A. Hooker, A. C. Michalos and J. W. van Evra (eds.), *PSA 1974. Proceedings of the 1974 Biennial Meeting of the Philosophy of Science Association.* [Boston Studies in the Philosophy of Science, Vol. XXXII] 1976 ISBN 90-277-0647-6; Pb 90-277-0648-4
102. Y. Fried and J. Agassi, *Paranoia.* A Study in Diagnosis. [Boston Studies in the Philosophy of Science, Vol. L] 1976 ISBN 90-277-0704-9; Pb 90-277-0705-7
103. M. Przełęcki, K. Szaniawski and R. Wójcicki (eds.), *Formal Methods in the Methodology of Empirical Sciences.* 1976 ISBN 90-277-0698-0
104. J. M. Vickers, *Belief and Probability.* 1976 ISBN 90-277-0744-8
105. K. H. Wolff, *Surrender and Catch.* Experience and Inquiry Today. [Boston Studies in the Philosophy of Science, Vol. LI] 1976 ISBN 90-277-0758-8; Pb 90-277-0765-0
106. K. Kosík, *Dialectics of the Concrete.* A Study on Problems of Man and World. [Boston Studies in the Philosophy of Science, Vol. LII] 1976 ISBN 90-277-0761-8; Pb 90-277-0764-2
107. N. Goodman, *The Structure of Appearance.* 3rd ed. with an Introduction by G. Hellman. [Boston Studies in the Philosophy of Science, Vol. LIII] 1977
ISBN 90-277-0773-1; Pb 90-277-0774-X
108. K. Ajdukiewicz, *The Scientific World-Perspective and Other Essays, 1931-1963.* Translated from Polish. Edited and with an Introduction by J. Giedymin. 1978 ISBN 90-277-0527-5
109. R. L. Causey, *Unity of Science.* 1977 ISBN 90-277-0779-0
110. R. E. Grandy, *Advanced Logic for Applications.* 1977 ISBN 90-277-0781-2
111. R. P. McArthur, *Tense Logic.* 1976 ISBN 90-277-0697-2
112. L. Lindahl, *Position and Change.* A Study in Law and Logic. Translated from Swedish by P. Needham. 1977 ISBN 90-277-0787-1
113. R. Tuomela, *Dispositions.* 1978 ISBN 90-277-0810-X
114. H. A. Simon, *Models of Discovery and Other Topics in the Methods of Science.* [Boston Studies in the Philosophy of Science, Vol. LIV] 1977 ISBN 90-277-0812-6; Pb 90-277-0858-4

SYNTHESE LIBRARY

115. R. D. Rosenkrantz, *Inference, Method and Decision*. Towards a Bayesian Philosophy of Science. 1977   ISBN 90-277-0817-7; Pb 90-277-0818-5
116. R. Tuomela, *Human Action and Its Explanation*. A Study on the Philosophical Foundations of Psychology. 1977   ISBN 90-277-0824-X
117. M. Lazerowitz, *The Language of Philosophy*. Freud and Wittgenstein. [Boston Studies in the Philosophy of Science, Vol. LV] 1977   ISBN 90-277-0826-6; Pb 90-277-0862-2
118. Not published 119. J. Pelc (ed.), *Semiotics in Poland, 1894–1969*. Translated from Polish. 1979   ISBN 90-277-0811-8
120. I. Pörn, *Action Theory and Social Science*. Some Formal Models. 1977 ISBN 90-277-0846-0
121. J. Margolis, *Persons and Mind*. The Prospects of Nonreductive Materialism. [Boston Studies in the Philosophy of Science, Vol. LVII] 1977   ISBN 90-277-0854-1; Pb 90-277-0863-0
122. J. Hintikka, I. Niiniluoto, and E. Saarinen (eds.), *Essays on Mathematical and Philosophical Logic*. 1979   ISBN 90-277-0879-7
123. T. A. F. Kuipers, *Studies in Inductive Probability and Rational Expectation*. 1978   ISBN 90-277-0882-7
124. E. Saarinen, R. Hilpinen, I. Niiniluoto and M. P. Hintikka (eds.), *Essays in Honour of Jaakko Hintikka on the Occasion of His 50th Birthday*. 1979   ISBN 90-277-0916-5
125. G. Radnitzky and G. Andersson (eds.), *Progress and Rationality in Science*. [Boston Studies in the Philosophy of Science, Vol. LVIII] 1978   ISBN 90-277-0921-1; Pb 90-277-0922-X
126. P. Mittelstaedt, *Quantum Logic*. 1978   ISBN 90-277-0925-4
127. K. A. Bowen, *Model Theory for Modal Logic*. Kripke Models for Modal Predicate Calculi. 1979   ISBN 90-277-0929-7
128. H. A. Bursen, *Dismantling the Memory Machine*. A Philosophical Investigation of Machine Theories of Memory. 1978   ISBN 90-277-0933-5
129. M. W. Wartofsky, *Models*. Representation and the Scientific Understanding. [Boston Studies in the Philosophy of Science, Vol. XLVIII] 1979   ISBN 90-277-0736-7; Pb 90-277-0947-5
130. D. Ihde, *Technics and Praxis*. A Philosophy of Technology. [Boston Studies in the Philosophy of Science, Vol. XXIV] 1979   ISBN 90-277-0953-X; Pb 90-277-0954-8
131. J. J. Wiatr (ed.), *Polish Essays in the Methodology of the Social Sciences*. [Boston Studies in the Philosophy of Science, Vol. XXIX] 1979   ISBN 90-277-0723-5; Pb 90-277-0956-4
132. W. C. Salmon (ed.), *Hans Reichenbach: Logical Empiricist*. 1979   ISBN 90-277-0958-0
133. P. Bieri, R.-P. Horstmann and L. Krüger (eds.), *Transcendental Arguments in Science*. Essays in Epistemology. 1979   ISBN 90-277-0963-7; Pb 90-277-0964-5
134. M. Marković and G. Petrović (eds.), *Praxis*. Yugoslav Essays in the Philosophy and Methodology of the Social Sciences. [Boston Studies in the Philosophy of Science, Vol. XXXVI] 1979   ISBN 90-277-0727-8; Pb 90-277-0968-8
135. R. Wójcicki, *Topics in the Formal Methodology of Empirical Sciences*. Translated from Polish. 1979   ISBN 90-277-1004-X
136. G. Radnitzky and G. Andersson (eds.), *The Structure and Development of Science*. [Boston Studies in the Philosophy of Science, Vol. LIX] 1979   ISBN 90-277-0994-7; Pb 90-277-0995-5
137. J. C. Webb, *Mechanism, Mentalism and Metamathematics*. An Essay on Finitism. 1980   ISBN 90-277-1046-5
138. D. F. Gustafson and B. L. Tapscott (eds.), *Body, Mind and Method*. Essays in Honor of Virgil C. Aldrich. 1979   ISBN 90-277-1013-9
139. L. Nowak, *The Structure of Idealization*. Towards a Systematic Interpretation of the Marxian Idea of Science. 1980   ISBN 90-277-1014-7

SYNTHESE LIBRARY

140. C. Perelman, *The New Rhetoric and the Humanities.* Essays on Rhetoric and Its Applications. Translated from French and German. With an Introduction by H. Zyskind. 1979
    ISBN 90-277-1018-X; Pb 90-277-1019-8
141. W. Rabinowicz, *Universalizability.* A Study in Morals and Metaphysics. 1979
    ISBN 90-277-1020-2
142. C. Perelman, *Justice, Law and Argument.* Essays on Moral and Legal Reasoning. Translated from French and German. With an Introduction by H.J. Berman. 1980
    ISBN 90-277-1089-9; Pb 90-277-1090-2
143. S. Kanger and S. Öhman (eds.), *Philosophy and Grammar.* Papers on the Occasion of the Quincentennial of Uppsala University. 1981   ISBN 90-277-1091-0
144. T. Pawlowski, *Concept Formation in the Humanities and the Social Sciences.* 1980
    ISBN 90-277-1096-1
145. J. Hintikka, D. Gruender and E. Agazzi (eds.), *Theory Change, Ancient Axiomatics and Galileo's Methodology.* Proceedings of the 1978 Pisa Conference on the History and Philosophy of Science, Volume I. 1981   ISBN 90-277-1126-7
146. J. Hintikka, D. Gruender and E. Agazzi (eds.), *Probabilistic Thinking, Thermodynamics, and the Interaction of the History and Philosophy of Science.* Proceedings of the 1978 Pisa Conference on the History and Philosophy of Science, Volume II. 1981   ISBN 90-277-1127-5
147. U. Mönnich (ed.), *Aspects of Philosophical Logic.* Some Logical Forays into Central Notions of Linguistics and Philosophy. 1981   ISBN 90-277-1201-8
148. D. M. Gabbay, *Semantical Investigations in Heyting's Intuitionistic Logic.* 1981
    ISBN 90-277-1202-6
149. E. Agazzi (ed.), *Modern Logic – A Survey.* Historical, Philosophical, and Mathematical Aspects of Modern Logic and Its Applications. 1981   ISBN 90-277-1137-2
150. A. F. Parker-Rhodes, *The Theory of Indistinguishables.* A Search for Explanatory Principles below the Level of Physics. 1981   ISBN 90-277-1214-X
151. J. C. Pitt, *Pictures, Images, and Conceptual Change.* An Analysis of Wilfrid Sellars' Philosophy of Science. 1981   ISBN 90-277-1276-X; Pb 90-277-1277-8
152. R. Hilpinen (ed.), *New Studies in Deontic Logic.* Norms, Actions, and the Foundations of Ethics. 1981   ISBN 90-277-1278-6; Pb 90-277-1346-4
153. C. Dilworth, *Scientific Progress.* A Study Concerning the Nature of the Relation between Successive Scientific Theories. 3rd rev. ed., 1994   ISBN 0-7923-2487-0; Pb 0-7923-2488-9
154. D. Woodruff Smith and R. McIntyre, *Husserl and Intentionality.* A Study of Mind, Meaning, and Language. 1982   ISBN 90-277-1392-8; Pb 90-277-1730-3
155. R. J. Nelson, *The Logic of Mind.* 2nd. ed., 1989   ISBN 90-277-2819-4; Pb 90-277-2822-4
156. J. F. A. K. van Benthem, *The Logic of Time.* A Model-Theoretic Investigation into the Varieties of Temporal Ontology, and Temporal Discourse. 1983; 2nd ed., 1991   ISBN 0-7923-1081-0
157. R. Swinburne (ed.), *Space, Time and Causality.* 1983   ISBN 90-277-1437-1
158. E. T. Jaynes, *Papers on Probability, Statistics and Statistical Physics.* Ed. by R. D. Rozenkrantz. 1983   ISBN 90-277-1448-7; Pb (1989) 0-7923-0213-3
159. T. Chapman, *Time: A Philosophical Analysis.* 1982   ISBN 90-277-1465-7
160. E. N. Zalta, *Abstract Objects.* An Introduction to Axiomatic Metaphysics. 1983
    ISBN 90-277-1474-6
161. S. Harding and M. B. Hintikka (eds.), *Discovering Reality.* Feminist Perspectives on Epistemology, Metaphysics, Methodology, and Philosophy of Science. 1983
    ISBN 90-277-1496-7; Pb 90-277-1538-6
162. M. A. Stewart (ed.), *Law, Morality and Rights.* 1983   ISBN 90-277-1519-X

SYNTHESE LIBRARY

163. D. Mayr and G. Süssmann (eds.), *Space, Time, and Mechanics*. Basic Structures of a Physical Theory. 1983 ISBN 90-277-1525-4
164. D. Gabbay and F. Guenthner (eds.), *Handbook of Philosophical Logic*. Vol. I: Elements of Classical Logic. 1983 ISBN 90-277-1542-4
165. D. Gabbay and F. Guenthner (eds.), *Handbook of Philosophical Logic*. Vol. II: Extensions of Classical Logic. 1984 ISBN 90-277-1604-8
166. D. Gabbay and F. Guenthner (eds.), *Handbook of Philosophical Logic*. Vol. III: Alternative to Classical Logic. 1986 ISBN 90-277-1605-6
167. D. Gabbay and F. Guenthner (eds.), *Handbook of Philosophical Logic*. Vol. IV: Topics in the Philosophy of Language. 1989 ISBN 90-277-1606-4
168. A. J. I. Jones, *Communication and Meaning*. An Essay in Applied Modal Logic. 1983
    ISBN 90-277-1543-2
169. M. Fitting, *Proof Methods for Modal and Intuitionistic Logics*. 1983 ISBN 90-277-1573-4
170. J. Margolis, *Culture and Cultural Entities*. Toward a New Unity of Science. 1984
    ISBN 90-277-1574-2
171. R. Tuomela, *A Theory of Social Action*. 1984 ISBN 90-277-1703-6
172. J. J. E. Gracia, E. Rabossi, E. Villanueva and M. Dascal (eds.), *Philosophical Analysis in Latin America*. 1984 ISBN 90-277-1749-4
173. P. Ziff, *Epistemic Analysis*. A Coherence Theory of Knowledge. 1984
    ISBN 90-277-1751-7
174. P. Ziff, *Antiaesthetics*. An Appreciation of the Cow with the Subtile Nose. 1984
    ISBN 90-277-1773-7
175. W. Balzer, D. A. Pearce, and H.-J. Schmidt (eds.), *Reduction in Science*. Structure, Examples, Philosophical Problems. 1984 ISBN 90-277-1811-3
176. A. Peczenik, L. Lindahl and B. van Roermund (eds.), *Theory of Legal Science*. Proceedings of the Conference on Legal Theory and Philosophy of Science (Lund, Sweden, December 1983). 1984 ISBN 90-277-1834-2
177. I. Niiniluoto, *Is Science Progressive?* 1984 ISBN 90-277-1835-0
178. B. K. Matilal and J. L. Shaw (eds.), *Analytical Philosophy in Comparative Perspective*. Exploratory Essays in Current Theories and Classical Indian Theories of Meaning and Reference. 1985 ISBN 90-277-1870-9
179. P. Kroes, *Time: Its Structure and Role in Physical Theories*. 1985 ISBN 90-277-1894-6
180. J. H. Fetzer, *Sociobiology and Epistemology*. 1985 ISBN 90-277-2005-3; Pb 90-277-2006-1
181. L. Haaparanta and J. Hintikka (eds.), *Frege Synthesized*. Essays on the Philosophical and Foundational Work of Gottlob Frege. 1986 ISBN 90-277-2126-2
182. M. Detlefsen, *Hilbert's Program*. An Essay on Mathematical Instrumentalism. 1986
    ISBN 90-277-2151-3
183. J. L. Golden and J. J. Pilotta (eds.), *Practical Reasoning in Human Affairs*. Studies in Honor of Chaim Perelman. 1986 ISBN 90-277-2255-2
184. H. Zandvoort, *Models of Scientific Development and the Case of Nuclear Magnetic Resonance*. 1986 ISBN 90-277-2351-6
185. I. Niiniluoto, *Truthlikeness*. 1987 ISBN 90-277-2354-0
186. W. Balzer, C. U. Moulines and J. D. Sneed, *An Architectonic for Science*. The Structuralist Program. 1987 ISBN 90-277-2403-2
187. D. Pearce, *Roads to Commensurability*. 1987 ISBN 90-277-2414-8
188. L. M. Vaina (ed.), *Matters of Intelligence*. Conceptual Structures in Cognitive Neuroscience. 1987 ISBN 90-277-2460-1

SYNTHESE LIBRARY

189. H. Siegel, *Relativism Refuted*. A Critique of Contemporary Epistemological Relativism. 1987
 ISBN 90-277-2469-5
190. W. Callebaut and R. Pinxten, *Evolutionary Epistemology*. A Multiparadigm Program, with a Complete Evolutionary Epistemology Bibliograph. 1987 ISBN 90-277-2582-9
191. J. Kmita, *Problems in Historical Epistemology*. 1988 ISBN 90-277-2199-8
192. J. H. Fetzer (ed.), *Probability and Causality*. Essays in Honor of Wesley C. Salmon, with an Annotated Bibliography. 1988 ISBN 90-277-2607-8; Pb 1-5560-8052-2
193. A. Donovan, L. Laudan and R. Laudan (eds.), *Scrutinizing Science*. Empirical Studies of Scientific Change. 1988 ISBN 90-277-2608-6
194. H.R. Otto and J.A. Tuedio (eds.), *Perspectives on Mind*. 1988 ISBN 90-277-2640-X
195. D. Batens and J.P. van Bendegem (eds.), *Theory and Experiment*. Recent Insights and New Perspectives on Their Relation. 1988 ISBN 90-277-2645-0
196. J. Österberg, *Self and Others*. A Study of Ethical Egoism. 1988 ISBN 90-277-2648-5
197. D.H. Helman (ed.), *Analogical Reasoning*. Perspectives of Artificial Intelligence, Cognitive Science, and Philosophy. 1988 ISBN 90-277-2711-2
198. J. Woleński, *Logic and Philosophy in the Lvov-Warsaw School*. 1989 ISBN 90-277-2749-X
199. R. Wójcicki, *Theory of Logical Calculi*. Basic Theory of Consequence Operations. 1988
 ISBN 90-277-2785-6
200. J. Hintikka and M.B. Hintikka, *The Logic of Epistemology and the Epistemology of Logic*. Selected Essays. 1989 ISBN 0-7923-0040-8; Pb 0-7923-0041-6
201. E. Agazzi (ed.), *Probability in the Sciences*. 1988 ISBN 90-277-2808-9
202. M. Meyer (ed.), *From Metaphysics to Rhetoric*. 1989 ISBN 90-277-2814-3
203. R.L. Tieszen, *Mathematical Intuition*. Phenomenology and Mathematical Knowledge. 1989
 ISBN 0-7923-0131-5
204. A. Melnick, *Space, Time, and Thought in Kant*. 1989 ISBN 0-7923-0135-8
205. D.W. Smith, *The Circle of Acquaintance*. Perception, Consciousness, and Empathy. 1989
 ISBN 0-7923-0252-4
206. M.H. Salmon (ed.), *The Philosophy of Logical Mechanism*. Essays in Honor of Arthur W. Burks. With his Responses, and with a Bibliography of Burk's Work. 1990
 ISBN 0-7923-0325-3
207. M. Kusch, *Language as Calculus vs. Language as Universal Medium*. A Study in Husserl, Heidegger, and Gadamer. 1989 ISBN 0-7923-0333-4
208. T.C. Meyering, *Historical Roots of Cognitive Science*. The Rise of a Cognitive Theory of Perception from Antiquity to the Nineteenth Century. 1989 ISBN 0-7923-0349-0
209. P. Kosso, *Observability and Observation in Physical Science*. 1989 ISBN 0-7923-0389-X
210. J. Kmita, *Essays on the Theory of Scientific Cognition*. 1990 ISBN 0-7923-0441-1
211. W. Sieg (ed.), *Acting and Reflecting*. The Interdisciplinary Turn in Philosophy. 1990
 ISBN 0-7923-0512-4
212. J. Karpiński, *Causality in Sociological Research*. 1990 ISBN 0-7923-0546-9
213. H.A. Lewis (ed.), *Peter Geach: Philosophical Encounters*. 1991 ISBN 0-7923-0823-9
214. M. Ter Hark, *Beyond the Inner and the Outer*. Wittgenstein's Philosophy of Psychology. 1990
 ISBN 0-7923-0850-6
215. M. Gosselin, *Nominalism and Contemporary Nominalism*. Ontological and Epistemological Implications of the Work of W.V.O. Quine and of N. Goodman. 1990 ISBN 0-7923-0904-9
216. J.H. Fetzer, D. Shatz and G. Schlesinger (eds.), *Definitions and Definability*. Philosophical Perspectives. 1991 ISBN 0-7923-1046-2
217. E. Agazzi and A. Cordero (eds.), *Philosophy and the Origin and Evolution of the Universe*. 1991 ISBN 0-7923-1322-4

SYNTHESE LIBRARY

218. M. Kusch, *Foucault's Strata and Fields*. An Investigation into Archaeological and Genealogical Science Studies. 1991 ISBN 0-7923-1462-X
219. C.J. Posy, *Kant's Philosophy of Mathematics*. Modern Essays. 1992 ISBN 0-7923-1495-6
220. G. Van de Vijver, *New Perspectives on Cybernetics*. Self-Organization, Autonomy and Connectionism. 1992 ISBN 0-7923-1519-7
221. J.C. Nyíri, *Tradition and Individuality*. Essays. 1992 ISBN 0-7923-1566-9
222. R. Howell, *Kant's Transcendental Deduction*. An Analysis of Main Themes in His Critical Philosophy. 1992 ISBN 0-7923-1571-5
223. A. García de la Sienra, *The Logical Foundations of the Marxian Theory of Value*. 1992
ISBN 0-7923-1778-5
224. D.S. Shwayder, *Statement and Referent*. An Inquiry into the Foundations of Our Conceptual Order. 1992 ISBN 0-7923-1803-X
225. M. Rosen, *Problems of the Hegelian Dialectic*. Dialectic Reconstructed as a Logic of Human Reality. 1993 ISBN 0-7923-2047-6
226. P. Suppes, *Models and Methods in the Philosophy of Science: Selected Essays*. 1993
ISBN 0-7923-2211-8
227. R. M. Dancy (ed.), *Kant and Critique: New Essays in Honor of W. H. Werkmeister*. 1993
ISBN 0-7923-2244-4
228. J. Woleński (ed.), *Philosophical Logic in Poland*. 1993 ISBN 0-7923-2293-2
229. M. De Rijke (ed.), *Diamonds and Defaults*. Studies in Pure and Applied Intensional Logic. 1993 ISBN 0-7923-2342-4
230. B.K. Matilal and A. Chakrabarti (eds.), *Knowing from Words*. Western and Indian Philosophical Analysis of Understanding and Testimony. 1994 ISBN 0-7923-2345-9
231. S.A. Kleiner, *The Logic of Discovery*. A Theory of the Rationality of Scientific Research. 1993
ISBN 0-7923-2371-8
232. R. Festa, *Optimum Inductive Methods*. A Study in Inductive Probability, Bayesian Statistics, and Verisimilitude. 1993 ISBN 0-7923-2460-9
233. P. Humphreys (ed.), *Patrick Suppes: Scientific Philosopher*. Vol. 1: Probability and Probabilistic Causality. 1994 ISBN 0-7923-2552-4
234. P. Humphreys (ed.), *Patrick Suppes: Scientific Philosopher*. Vol. 2: Philosophy of Physics, Theory Structure, and Measurement Theory. 1994 ISBN 0-7923-2553-2
235. P. Humphreys (ed.), *Patrick Suppes: Scientific Philosopher*. Vol. 3: Language, Logic, and Psychology. 1994 ISBN 0-7923-2862-0
Set ISBN (Vols 233–235) 0-7923-2554-0
236. D. Prawitz and D. Westerståhl (eds.), *Logic and Philosophy of Science in Uppsala*. Papers from the 9th International Congress of Logic, Methodology, and Philosophy of Science. 1994
ISBN 0-7923-2702-0
237. L. Haaparanta (ed.), *Mind, Meaning and Mathematics*. Essays on the Philosophical Views of Husserl and Frege. 1994 ISBN 0-7923-2703-9
238. J. Hintikka (ed.), *Aspects of Metaphor*. 1994 ISBN 0-7923-2786-1
239. B. McGuinness and G. Oliveri (eds.), *The Philosophy of Michael Dummett*. With Replies from Michael Dummett. 1994 ISBN 0-7923-2804-3
240. D. Jamieson (ed.), *Language, Mind, and Art*. Essays in Appreciation and Analysis, In Honor of Paul Ziff. 1994 ISBN 0-7923-2810-8
241. G. Preyer, F. Siebelt and A. Ulfig (eds.), *Language, Mind and Epistemology*. On Donald Davidson's Philosophy. 1994 ISBN 0-7923-2811-6
242. P. Ehrlich (ed.), *Real Numbers, Generalizations of the Reals, and Theories of Continua*. 1994
ISBN 0-7923-2689-X

SYNTHESE LIBRARY

243. G. Debrock and M. Hulswit (eds.), *Living Doubt.* Essays concerning the epistemology of Charles Sanders Peirce. 1994 ISBN 0-7923-2898-1
244. J. Srzednicki, *To Know or Not to Know.* Beyond Realism and Anti-Realism. 1994
 ISBN 0-7923-2909-0
245. R. Egidi (ed.), *Wittgenstein: Mind and Language.* 1995 ISBN 0-7923-3171-0
246. A. Hyslop, *Other Minds.* 1995 ISBN 0-7923-3245-8
247. L. Pólos and M. Masuch (eds.), *Applied Logic: How, What and Why.* Logical Approaches to Natural Language. 1995 ISBN 0-7923-3432-9
248. M. Krynicki, M. Mostowski and L.M. Szczerba (eds.), *Quantifiers: Logics, Models and Computation.* Volume One: Surveys. 1995 ISBN 0-7923-3448-5
249. M. Krynicki, M. Mostowski and L.M. Szczerba (eds.), *Quantifiers: Logics, Models and Computation.* Volume Two: Contributions. 1995 ISBN 0-7923-3449-3
 Set ISBN (Vols 248 + 249) 0-7923-3450-7
250. R.A. Watson, *Representational Ideas from Plato to Patricia Churchland.* 1995
 ISBN 0-7923-3453-1
251. J. Hintikka (ed.), *From Dedekind to Gödel.* Essays on the Development of the Foundations of Mathematics. 1995 ISBN 0-7923-3484-1
252. A. Wiśniewski, *The Posing of Questions.* Logical Foundations of Erotetic Inferences. 1995
 ISBN 0-7923-3637-2
253. J. Peregrin, *Doing Worlds with Words.* Formal Semantics without Formal Metaphysics. 1995
 ISBN 0-7923-3742-5
254. I.A. Kieseppä, *Truthlikeness for Multidimensional, Quantitative Cognitive Problems.* 1996
 ISBN 0-7923-4005-1
255. P. Hugly and C. Sayward: *Intensionality and Truth.* An Essay on the Philosophy of A.N. Prior. 1996 ISBN 0-7923-4119-8
256. L. Hankinson Nelson and J. Nelson (eds.): *Feminism, Science, and the Philosophy of Science.* 1997 ISBN 0-7923-4162-7
257. P.I. Bystrov and V.N. Sadovsky (eds.): *Philosophical Logic and Logical Philosophy.* Essays in Honour of Vladimir A. Smirnov. 1996 ISBN 0-7923-4270-4
258. Å.E. Andersson and N-E. Sahlin (eds.): *The Complexity of Creativity.* 1996
 ISBN 0-7923-4346-8
259. M.L. Dalla Chiara, K. Doets, D. Mundici and J. van Benthem (eds.): *Logic and Scientific Methods.* Volume One of the Tenth International Congress of Logic, Methodology and Philosophy of Science, Florence, August 1995. 1997 ISBN 0-7923-4383-2
260. M.L. Dalla Chiara, K. Doets, D. Mundici and J. van Benthem (eds.): *Structures and Norms in Science.* Volume Two of the Tenth International Congress of Logic, Methodology and Philosophy of Science, Florence, August 1995. 1997 ISBN 0-7923-4384-0
 Set ISBN (Vols 259 + 260) 0-7923-4385-9
261. A. Chakrabarti: *Denying Existence.* The Logic, Epistemology and Pragmatics of Negative Existentials and Fictional Discourse. 1997 ISBN 0-7923-4388-3
262. A. Biletzki: *Talking Wolves.* Thomas Hobbes on the Language of Politics and the Politics of Language. 1997 ISBN 0-7923-4425-1
263. D. Nute (ed.): *Defeasible Deontic Logic.* 1997 ISBN 0-7923-4630-0
264. U. Meixner: *Axiomatic Formal Ontology.* 1997 ISBN 0-7923-4747-X
265. I. Brinck: *The Indexical 'I'.* The First Person in Thought and Language. 1997
 ISBN 0-7923-4741-2
266. G. Hölmström-Hintikka and R. Tuomela (eds.): *Contemporary Action Theory.* Volume 1: Individual Action. 1997 ISBN 0-7923-4753-6; Set: 0-7923-4754-4

SYNTHESE LIBRARY

267. G. Hölmström-Hintikka and R. Tuomela (eds.): *Contemporary Action Theory.* Volume 2: Social Action. 1997   ISBN 0-7923-4752-8; Set: 0-7923-4754-4
268. B.-C. Park: *Phenomenological Aspects of Wittgenstein's Philosophy.* 1998
 ISBN 0-7923-4813-3
269. J. Paśniczek: *The Logic of Intentional Objects.* A Meinongian Version of Classical Logic. 1998
 Hb ISBN 0-7923-4880-X; Pb ISBN 0-7923-5578-4
270. P.W. Humphreys and J.H. Fetzer (eds.): *The New Theory of Reference.* Kripke, Marcus, and Its Origins. 1998   ISBN 0-7923-4898-2
271. K. Szaniawski, A. Chmielewski and J. Woleński (eds.): *On Science, Inference, Information and Decision Making.* Selected Essays in the Philosophy of Science. 1998
 ISBN 0-7923-4922-9
272. G.H. von Wright: *In the Shadow of Descartes.* Essays in the Philosophy of Mind. 1998
 ISBN 0-7923-4992-X
273. K. Kijania-Placek and J. Woleński (eds.): *The Lvov–Warsaw School and Contemporary Philosophy.* 1998   ISBN 0-7923-5105-3
274. D. Dedrick: *Naming the Rainbow.* Colour Language, Colour Science, and Culture. 1998
 ISBN 0-7923-5239-4
275. L. Albertazzi (ed.): *Shapes of Forms.* From Gestalt Psychology and Phenomenology to Ontology and Mathematics. 1999   ISBN 0-7923-5246-7
276. P. Fletcher: *Truth, Proof and Infinity.* A Theory of Constructions and Constructive Reasoning. 1998   ISBN 0-7923-5262-9
277. M. Fitting and R.L. Mendelsohn (eds.): *First-Order Modal Logic.* 1998
 Hb ISBN 0-7923-5334-X; Pb ISBN 0-7923-5335-8
278. J.N. Mohanty: *Logic, Truth and the Modalities from a Phenomenological Perspective.* 1999
 ISBN 0-7923-5550-4
279. T. Placek: *Mathematical Intiutionism and Intersubjectivity.* A Critical Exposition of Arguments for Intuitionism. 1999   ISBN 0-7923-5630-6
280. A. Cantini, E. Casari and P. Minari (eds.): *Logic and Foundations of Mathematics.* 1999
 ISBN 0-7923-5659-4   set ISBN 0-7923-5867-8
281. M.L. Dalla Chiara, R. Giuntini and F. Laudisa (eds.): *Language, Quantum, Music.* 1999
 ISBN 0-7923-5727-2; set ISBN 0-7923-5867-8
282. R. Egidi (ed.): *In Search of a New Humanism.* The Philosophy of Georg Hendrik von Wright. 1999   ISBN 0-7923-5810-4
283. F. Vollmer: *Agent Causality.* 1999   ISBN 0-7923-5848-1
284. J. Peregrin (ed.): *Truth and Its Nature (if Any).* 1999   ISBN 0-7923-5865-1
285. M. De Caro (ed.): *Interpretations and Causes.* New Perspectives on Donald Davidson's Philosophy. 1999   ISBN 0-7923-5869-4
286. R. Murawski: *Recursive Functions and Metamathematics.* Problems of Completeness and Decidability, Gödel's Theorems. 1999   ISBN 0-7923-5904-6
287. T.A.F. Kuipers: *From Instrumentalism to Constructive Realism.* On Some Relations between Confirmation, Empirical Progress, and Truth Approximation. 2000   ISBN 0-7923-6086-9
288. G. Holmström-Hintikka (ed.): *Medieval Philosophy and Modern Times.* 2000
 ISBN 0-7923-6102-4
289. E. Grosholz and H. Breger (eds.): *The Growth of Mathematical Knowledge.* 2000
 ISBN 0-7923-6151-2

SYNTHESE LIBRARY

290. G. Sommaruga: *History and Philosophy of Constructive Type Theory.* 2000
ISBN 0-7923-6180-6
291. J. Gasser (ed.): *A Boole Anthology.* Recent and Classical Studies in the Logic of George Boole. 2000
ISBN 0-7923-6380-9
292. V.F. Hendricks, S.A. Pedersen and K.F. Jørgensen (eds.): *Proof Theory.* History and Philosophical Significance. 2000
ISBN 0-7923-6544-5
293. W.L. Craig: *The Tensed Theory of Time.* A Critical Examination. 2000   ISBN 0-7923-6634-4
294. W.L. Craig: *The Tenseless Theory of Time.* A Critical Examination. 2000
ISBN 0-7923-6635-2
295. L. Albertazzi (ed.): *The Dawn of Cognitive Science.* Early European Contributors. 2001
ISBN 0-7923-6799-5
296. G. Forrai: *Reference, Truth and Conceptual Schemes.* A Defense of Internal Realism. 2001
ISBN 0-7923-6885-1
297. V.F. Hendricks, S.A. Pedersen and K.F. Jørgensen (eds.): *Probability Theory.* Philosophy, Recent History and Relations to Science. 2001
ISBN 0-7923-6952-1

Previous volumes are still available.

KLUWER ACADEMIC PUBLISHERS – DORDRECHT / BOSTON / LONDON

OHIO UNIVERSITY LIBRARY
Please return this book as soon as you have
finished with it. In order to avoid a fine it must